Lecture Notes in Mathematics

1556

Editors:
A. Dold, Heidelberg
B. Eckmann, Zürich
F. Takens, Groningen

Sergej B. Kuksin

Nearly Integrable Infinite-Dimensional Hamiltonian Systems

Springer-Verlag
Berlin Heidelberg New York
London Paris Tokyo
Hong Kong Barcelona
Budapest

Author

Sergej B. Kuksin
Institute for Information
Transmission Problems
Ermolovoy St. 19
101 447 Moscow, Russia

Mathematics Subject Classification (1991): 34G20, 35Q55, 58F07, 58F27, 58F39

ISBN 3-540-57161-2 Springer-Verlag Berlin Heidelberg New York
ISBN 0-387-57161-2 Springer-Verlag New York Berlin Heidelberg

Printing and binding: Druckhaus Beltz, Hemsbach/Bergstr.
2146/3140-543210 - Printed on acid-free paper

Contents

Introduction

The book is devoted to nonlinear Hamiltonian perturbations of integrable (linear and nonlinear) Hamiltonian systems of large and infinite dimension. Such systems arise in physics in many different ways. As a working hypothesis for theirs study it was postulated in the physical literature after the works of Boltzmann that in a "typical situation" their solutions are stochastic. This postulate ("ergodic hypothesis") was successfully used to explain many properties of matter. On the other hand, a lot of numerical experiments starting from the ones of Fermi–Pasta–Ulam (see [FPU], [U]) have shown quite regular recurrent behavior of many solutions of the systems under consideration (see e.g. [ZIS]). This effect cannot be explained by means of the Poincaré recurrence theorem [AKN] because the Poincaré recurrence time is much larger than the one obtained in the experiments. It seems that the investigated systems have in abundance quasiperiodic trajectories or trajectories abnormally close to the quasiperiodic ones (see [LL], [DEGM], [Mo]). These trajectories correspond to low-frequency oscillations of the underlying physical object. In these oscillations the energy is frozen in low frequencies for a very long time. So the recurrence effect causes a low rate of stochasticity (the ergodic hypothesis works now in a slow way). This effect seemed rather strange to the physicists who observed it.

Our goal in this book is to obtain some general theorem to prove that "many" quasiperiodic solutions of the unperturbed integrable system, which describes a conservative physical system with one spatial dimension, persist under perturbations. The theorem gives some explanation to the recurrence effect in spatially one-dimensional systems. It proves that in some strict sense the *one-dimensional world* "is not very ergodic".

We consider *discrete*-spectrum systems only. For Hamiltonian systems with *continuous* spectrum time-quasiperiodic solutions play rather unessential role. To study nearintegrable continuous-spectrum systems various types of averaging theorems in time- and space-variables have been developed. We avoid discussing of this expanded subject.

The main part of the book deals with perturbations of *linear* Hamiltonian equations, depending on a finite-dimensional parameter. However, it turns out that the problem of persistence quasiperiodic solutions of a *nonlinear* integrable system can be reduced to the same problem for a parameter-depending linear equation (after the reduction the frequency vector of the unperturbed nonlinear quasiperiodic oscillation plays a role of the parameter we need). Similar finite-dimensional reduction is well-known; see Remark in item 1.3 below. We discuss the infinite-dimensional case in item 3.2 of the introduction and refer the reader for details to original papers (and, hopefully, to the next book of the author). We formulate the main theorem of the book in a way to simplify its nonlinear applications. Therefore the title of the book refer to *integrable* (linear and nonlinear) systems rather than to *linear* systems only.

The introduction is devoted to a rather expanded discussion of the theorem and its applications. Sometimes the discussion supplements the results from the main text. We preface the survey of our results with a survey of the finite dimensional situation.

1. Finite dimensional situation

"Regular" (periodic and quasiperiodic) solutions of $2n$-dimensional Hamiltonian systems are important for classical and celestial mechanics. Some quite general existence theorems for this class of solutions have been obtained. Here we are interested in perturbation-type results only.

1.1 Lyapunov and Poincaré theorems

The first classical results in this direction where the Lyapunov and Poincaré theorems (see [AKN], [SM]), stating that nonresonant periodic solutions of a Hamiltonian system survive under Hamiltonian perturbation. More exactly, the Lyapunov theorem states that if the unperturbed system is a linear Hamiltonian system with the spectrum

$$\{\pm i\lambda\} \cup \{\pm\mu_1, \dots, \pm\mu_{n-1}\},$$

where $\lambda \in \mathbb{R}\backslash\{0\}$, $\mu_1, \dots, \mu_{n-1}$ are complex numbers and

$$ik\lambda \neq \mu_j \quad \forall j = 1, \dots, n-1, \quad \forall k \in \mathbb{Z},$$

then the perturbed system has a two-dimensional invariant manifold filled with periodic solutions of frequencies close to λ (i.e., of periods close to $2\pi/\lambda$).

The Poincaré theorem states that if a Hamiltonian system has a periodic solution such that the linearization of the corresponding isoenergetic Poincaré map at the fixed point does not have an eigenvalue equal to one, then this solution lies in two-dimensional invariant manifold filled with periodic solutions. The periodic solutions and the manifold they fill persist under Hamiltonian perturbations of the equation.

1.2 Kolmogorov theorem

The second classical result concerning the subject is the Kolmogorov theorem [Kol] (stated in [Kol] with a scheme of a proof given, and proven in details by Arnold and Moser), which inspired Arnold and Moser to create a powerful technique to handle nonlinear problems, well known nowadays as KAM (Kolmogorov–Arnold–Moser) theory; see [A2], [A3], [AA], [Mo], [SM] and bibliographies of the last three books. Kolmogorov's theorem states that most of the quasiperiodic n-frequency solutions of a nondegenerate integrable analytical system with n degrees of freedom persist under analytic Hamiltonian perturbations or, equivalently, Hamiltonian perturbations preserve most of invariant n-tori of a nondegenerate integrable system. Here integrability means that in a phase space $\mathbb{T}^n \times P$ (P is a bounded n-dimensional domain) the system has the form:

$$\dot{q} = \nabla h(p), \quad \dot{p} = 0, \tag{2}$$

(i.e., it has a hamiltonian h depending on the actions $p \in P$ only) and the nondegeneracy means that

$$\operatorname{Hess} h(p) := \det\{\partial^2 h(p)/\partial p_i \partial p_j\} \neq 0 . \tag{3}$$

Invariant tori of the system (2) are of the form

$$T^n(p) = \mathrm{T}^n \times \{p\} , \quad p \in P , \tag{4}$$

and most of them survive in the perturbed system with the hamiltonian $h(p) + \varepsilon H(q,p)$,

$$\dot{q} = \nabla_p\big(h(p) + \varepsilon H(q,p)\big) , \quad \dot{p} = -\varepsilon \nabla_q H(q,p) , \tag{5}$$

if positive ε is small enough. That means that for $\rho < 1$ there exists a subset $P_\varepsilon \subset P$ such that $\operatorname{mes}(P \backslash P_\varepsilon) \to 0$ as $\varepsilon \to 0$, and for $p \in P_\varepsilon$ there exists a map $\Sigma_p : \mathrm{T}^n \to \mathrm{T}^n \times P$ and an n-vector $\omega(p)$ such that $|\omega(p) - \nabla h(p)| \leq C\varepsilon$, for all $q \in \mathrm{T}^n$ $\operatorname{dist}\big(\Sigma_p(q),(q,p)\big) < \varepsilon^\rho$ and the curve

$$t \longmapsto \Sigma_p\big(q + t\omega(p)\big) \tag{6}$$

is a solution of (5).

For other versions and important improvements of the theorem se [AKN], [Bru1], [Bru2], [Her], [Laz], [Mo], [Mo1], [P3], [Ru], [Sev], [SZ], [Z1].

1.3 Melnikov theorem

The Lyapunov and Poincaré theorems state the persistence of nondegenerate one-dimensional invariant tori (= periodic solutions) under Hamiltonian perturbations, and the Kolmogorov theorem states the persistence most of the invariant N-tori of integrable system with N degrees of freedom. The natural question is if most of invariant tori of an intermediate dimension n, $1 < n < N$, survive under perturbations. For perturbations of a linear Hamiltonian system with $N = n + m$ degrees of freedom the question means the following. In the phase space

$$\mathrm{T}^n \times P \times \mathrm{R}^{2m} = \{(q,p,z)\} , \quad z = (z_+, z_-) \in \mathrm{R}^{2m} , \tag{7}$$

($n \geq 2, m \geq 1$) the Hamiltonian equations

$$\dot{q} = \lambda + \varepsilon \nabla_p H , \quad \dot{p} = -\varepsilon \nabla_q H , \quad \dot{z} = J(Az + \varepsilon \nabla_z H) , \tag{8}$$

are considered. Here $J(z_+, z_-) = (-z_-, z_+)$, A is a symmetric linear operator in R^{2m}, $\varepsilon H = \varepsilon H(q,p,z)$ is an analytic perturbation and $\lambda \in \Lambda \subset \mathrm{R}^n$ is a parameter. For $\varepsilon = 0$ the system (8) has invariant n-tori $T^{n,m}(p) = \mathrm{T}^n \times \{p\} \times \{0\}$, $p \in P$. The question is if these tori persist in the system (8) for $\varepsilon > 0$.

Let us denote the spectrum of the operator JA by $M = \{\mu_1, \ldots, \mu_{2m}\}$. We should distinguish three cases:

a) (*nondegenerate hyperbolic tori*) $M \subset C \backslash i\mathrm{R}$, $\mu_j \neq \mu_k$ $\forall_j \neq k$. In this situation a hyperbolic torus $T^{n,m}(p)$ persist for most λ. That is, for positive ε small enough and for $\lambda \in \Lambda(\varepsilon, p)$, where $\operatorname{mes} \Lambda \setminus \Lambda(\varepsilon, p) \to 0$ $(\varepsilon \to 0)$, the equation (8) has an invariant torus at a distance $< \varepsilon^p$ from $T^{n,m}(p)$. See [Gr], [Mo], [Z1].

b$_0$) (*nondegenerate elliptic tori*) $M \subset i\mathbb{R}\setminus\{0\}$, $\mu_j \neq \mu_k$ $\forall j \neq k$. This situation is more complicated. The preservation theorem for the elliptic torus $T^{n,m}(p)$ for most λ was formulated by Melnikov [Me1], [Me2]. The complete proof of the theorem was published only 15 years later by Eliasson [El], Pöschel [P1] and the author [K1], [K2] (the infinite-dimensional theorems of the last two works are applicable to equations (8) as well). The proofs given in the papers just mentioned are also valid in the more general situation:

b) (*nondegenerate tori*) $0 \notin M$, $\mu_j \neq \mu_k$ $\forall j \neq k$.

In the degenerate case

c) $0 \in M$ or $\mu_j = \mu_k$ for some $j \neq k$

no preservation theorem for the tori $T^{n,m}(p)$, formulated in terms of the unperturbed equation (8) with $\varepsilon = 0$ only, is known yet.

Remark. Melnikov theorem (case b$_0$)) remains true for $m = 0$, too. In such a case $Y = \{0\}$ and the theorem asserts the preservation of the n-dimensional invariant torus $\mathrm{T}^n \times \{0\}$ of the system with the linear hamiltonian $h(p) = \lambda \cdot p$,

$$\dot{q} = \lambda\,, \quad \dot{p} = 0\,,$$

under small analytic Hamiltonian perturbations for most parameters $\lambda \in \Lambda$. This result implies Kolmogorov's theorem as it was formulated above via some simple substitution; see [Mo1], p. 171.[1] Conversely, one can easily extract somewhat different version of Melnikov's theorem with $m = 0$ from Kolmogorov's theorem.[2] So these two statement are essentially equivalent. This equivalence (we had found it in the paper [Mo1]) was important for our insight into infinite-dimensional problems.

Remark. The equation (8) arises in studies of nearintegrable systems (5) with $n := N$ near the tori (4) "with some cycles shrinked to zero". It means that we suppose the system (5) be in the Birkhoff normal form (i.e., in the phase-space $\mathbb{R}^N_x \times \mathbb{R}^N_y$ with the usual symplectic structure it has an analytic hamiltonian $h(x_1^2 + y_1^2, \ldots, x_N^2 + y_N^2)$) and study its perturbations near an invariant n-torus $\{x_j^2 + y_j^2 = 2I_j\}$, where $I_1, \ldots, I_n > 0 = I_{n+1} = \cdots = I_N$.

Lower-dimensional invariant tori also fill resonant Lagrangian (= half-dimensional) tori (4), but they always lead to the degenerate case c). In particular, if for

---

1) In (5) substitute $p = a + \sqrt{\varepsilon}\tilde{p}$, $q = \tilde{q}$, regarding $a \in P$ as a parameter of the substitution. In the tilde-variables the hamiltonian $h + \varepsilon H$ equals const $+ \nabla h(a) \cdot \tilde{p} + O(\sqrt{\varepsilon})$, and we get the system (8) with $\varepsilon := \sqrt{\varepsilon}$, $m = 0$, $\lambda = \nabla h(a)$. As $\operatorname{Hess} h \not\equiv 0$, then we can treat $\lambda \in \nabla h(P)$ as a new parameter and apply Melnikov theorem.

2) Given a system (8) with $m = 0$, consider the extended phase-space $(\mathrm{T}^n \times P) \times (\mathrm{T}^n \times \Lambda) = \{(q, p, \tilde{q}, \tilde{p})\}$ and the hamiltonian $\tilde{H}_\varepsilon = p \cdot \tilde{p} + \varepsilon H(q, p)$. The nondegeneracy assumption (3) holds for the function $(p, \tilde{p}) \longmapsto p \cdot \tilde{p}$ and Kolmogorov theorem (with $n := 2n$) can be aplied. The invariant tori of the hamiltonian $\tilde{H}_\varepsilon$ have the form $T^n_\varepsilon \times (\mathrm{T}^n \times \{\tilde{p}\})$, where $\tilde{p} \in \Lambda$ and $T^n_\varepsilon \subset \mathrm{T}^n \times P$ is an invariant torus of (8).

$E(p) := \mathbb{Q}\partial h/\partial p_1 + \cdots + \mathbb{Q}\partial h/\partial p_N$ we have $\dim_{\mathbb{Q}} E(p) = N-1$, then the torus $T^N(p)$ is filled with invariant $(N-1)$-tori. Near each $(N-1)$-torus the perturbed system (5) may be reduced to a system (8) with JA equal to the Jordan 2×2-cell with zero eigenvalue. For more information see [Lo] and references therein.

Famous finite-gap time-quasiperiodic solutions of integrable PDE's form finite-dimensional invariant tori of the corresponding infinite-dimensional Hamiltonian integrable systems, obtained by "shrinking" (not "degenerating" !) of half-dimensional invariant tori. See below and [McT], [K7].

2. Infinite dimensional systems

2.1 The Problem

In a Hilbert space Z with inner product $\langle\cdot,\cdot\rangle$ we consider the equation

$$\dot u(t) = J\nabla\mathcal{K}\big(u(t)\big)\ , \quad u(t)\in Z\ . \tag{9}$$

Here J is an antiselfadjoint operator in Z and $\nabla\mathcal{K}$ is the gradient of a functional $\mathcal{K}$ relative to the inner product $\langle\cdot,\cdot\rangle$. In the most interesting situations the linear operator J, or the nonlinear operator $\nabla\mathcal{K}$, or both of them are unbounded. So one has to be careful with the equation and its solutions. For the exact definition of solutions of (9) and for some their properties see [Bre], [Lio] and Part 1 of the main text. Equation (9) is Hamiltonian if the phase space Z is provided with a symplectic structure by means of 2-form $-\langle J^{-1}du, du\rangle$ (by definition, $-\langle J^{-1}du,du\rangle[\xi,\eta] = -\langle J^{-1}\xi,\eta\rangle$ for ξ,η in Z).

In this book we are most interested in equations of the form

$$\dot u(t) = J\Big(Au(t) + \varepsilon\nabla H\big(u(t)\big)\Big)\ . \tag{10}$$

This equation is Hamiltonian with the hamiltonian

$$\mathcal{K}_\varepsilon = \frac{1}{2}\langle Au,u\rangle + \varepsilon H(u)\ .$$

Here A is a selfadjoint linear operator in Z and H is an analytic functional. The linear operators J, A and the nonlinear operator ∇H are assumed to be characterized by their *orders* d^J, d^A and d^H. We suppose that

$$d^J\geq 0\ ,\ d^A\geq 0\ ,\ d^J+d^A\geq 1\ ,\ d^J+d^H\leq 0\ . \tag{11}$$

In the most important examples Z is the L_2-space of square-summable functions on a segment, and J and A are differential operators. In such a case d^J, d^A are the orders of the differential operators and $\nabla H(u)$ is a variational derivative $\delta H/\delta u(x)$. In particular, if

$$H(u) = \int h(x,u(x))\,dx\ ,$$

then $\delta H/\delta u(x) = h_u\big(x,u(x)\big)$ and $d^H=0$; if the density h depends on integral of $u(x)$ instead of $u(x)$ itself, then $d^H<0$. To define the orders d^J, d^A, d^H in a general case, we include the space Z into a scale of Hilbert spaces. See Part 1 below.

The assumption (11) implies that equation (10) is quasilinear. This assumption is rather natural for the study of long-time behavior of solutions because for some strongly nonlinear Hamiltonian equations (i.e. ones of the form (10) with $d^H = d^A$) it is known that the equations have no nontrivial solutions existing for all time; see [Lax].

We suppose that J and A commute and that Z admits an orthonormal basis $\{\varphi_j^{\pm} \mid j \geq 1\}$ such that

$$A\varphi_j^{\pm} = \lambda_j^A \varphi_j^{\pm}\,, \quad J\varphi_j^{\pm} = \mp\lambda_j^J \varphi_j^{\mp}\,, \quad \forall j \geq 1\,. \tag{12}$$

So, in particular, the spectrum of the operator JA is equal to

$$\{\pm i\lambda_j \mid j \geq 1, \lambda_j = \lambda_j^A \lambda_j^J\}\,.$$ [3)]

Let us fix some $n \geq 1$. The $2n$-dimensional linear space

$$Z^0 = \operatorname{span}\{\varphi_j^{\pm} \mid 1 \leq j \leq n\}$$

is invariant for the flow of equation (10) with $\varepsilon = 0$, is foliated into invariant n-tori

$$T^n(I) = \{\sum_{j=1}^{n} x_j^{\pm}\varphi_j^{\pm} \mid {x_j^+}^2 + {x_j^-}^2 = 2I_j \ \forall j\}\,,$$

$I = (I_1, \ldots, I_n) \in \mathbb{R}_+^n$, and every torus $T^n(I)$ is filled with quasiperiodic solutions of the equation.[4)]

One can treat (10) with $\varepsilon = 0$ as an infinite chain of free harmonic oscillators with the frequencies $\lambda_1, \lambda_2, \ldots$. The solutions lying on the tori $T^n(I)$ correspond to oscillations with only the first n oscillators being excited. One can treat these solutions as low-frequency oscillations.

We study the question: *under what assumptions do the tori $T^n(I)$ and the corresponding low-frequency quasiperiodic solutions persist in equation (10) for $\varepsilon > 0$?*

It is convenient to introduce the angle-action variables $(q_1, \ldots, q_n, p_1, \ldots, p_n)$ in the space Z^0,

$$x_j^+ + ix_j^- = \sqrt{2p_j}\exp(iq_j)\,, \quad j = 1, \ldots, n$$

($x_j^{\pm}$ are the coordinates with respect to the basis $\{\varphi_j^{\pm} \mid 1 \leq j \leq n\}$); to denote by $Y = Z \ominus Z^0$ the closure of $\operatorname{span}\{\varphi_j^{\pm} \mid j \geq n+1\}$ and to pass to the variables (q, p, y),

$$q = (q_1, \ldots, q_n) \in \mathbb{T}^n\,, \quad p = (p_1, \ldots, p_n) \in \mathbb{R}_+^n\,, \quad y \in Y\,. \tag{13}$$

[3)] The assumption (12) may be essentially weakened. See Part 2.7.

[4)] We remain that a solution $u(t)$ is called *quasiperiodic* with n frequencies if there exist a continuous map $U : \mathbb{T}^n \to Z$ and n-vector ω (called the *frequency vector* of the solution) such that $u(t) \equiv U(\omega t)$. A quasiperiodic solution with one frequency is periodic, so quasiperiodic solutions represent a natural extension of the class of periodic solutions.

Let us denote by Σ^0 the imbedding

$$\Sigma^0 : \mathrm{T}^n \times \mathrm{R}^N_+ \longrightarrow Z\ , \quad (q,p) \longmapsto (q,p,0)\ .$$

(we use in Z the coordinates (13)). The invariant space Z^0 is the image of this map.

In the new variables (13) equation (10) takes the form:

$$\dot q = \nabla_p \mathcal{H}\ , \quad \dot p = -\nabla_q \mathcal{H}\ , \quad \dot y = J^Y \nabla_y \mathcal{H} \tag{14}$$

with

$$\mathcal{H} = \mathcal{H}_\varepsilon = \omega \cdot p + \frac{1}{2}\langle A^Y y, y\rangle + \varepsilon H(q,p,y)\ .$$

Here $\omega = (\lambda_1, \ldots, \lambda_n)$, $J^Y = J_{|Y}$, $A^Y = A_{|Y}$. So the operator $J^Y A^Y$ has pure imaginary spectrum $\{\pm i\lambda_j \mid j \geq n+1\}$ and one can easily recognize in the last equations an infinite-dimensional analogy to the elliptic case of the system (8). The form of Melnikov's theorem we gave above in Section 1.3 has a natural infinite-dimensional reformulation. It is remarkable that this reformulation becomes a true statement after adding essentially just two infinite-dimensional conditions.

2.2 The result

Keeping in mind the applications, we suppose that equation (10) analytically depends on n outer parameters $(a_1, \ldots, a_n) = a \in \mathfrak{A}$, where $\mathfrak{A}$ is a connected bounded open domain in R^n. So $A = A_a$, $H = H_a$ and $\lambda_j = \lambda_j(a)$. Let us assume that

$$\det\{\partial\lambda_j(a)/\partial a_k \mid 1 \leq j,k \leq n\} \not\equiv 0\ . \tag{15}$$

This assumption means that we can replace the parameter a by $\omega = (\lambda_1, \ldots, \lambda_n)(a)$. Later we refer to ω as to the *natural parameter* of the equation.

We consider a torus $T^n(I_1, \ldots, I_n)$ such that $I_k > 0\ \forall k$.

Theorem 1. Let us suppose that the assumptions (12), (15) hold together with

1) *(quasilinearity)*

$$d^J \geq 0\ , \quad d^A \geq 0\ , \quad d_1 := d^J + d^A \geq 1\ , \quad d^J + d^H \leq 0\ , \quad d^J + d^H < d_1 - 1\ ,$$

2) *(spectral asymptotics)*

$$\lambda_j(a) = K_1 j^{d_1} + K_2 + \mu_j(a)\ ,$$

where

$$|\mu_j(a)| + |\nabla\mu_j(a)| \leq K_3 j^{d_1 - \kappa}$$

for some $\kappa > 1$;

3) for some $N \geq n$ and $M \geq 1$ depending on the problem (10) the *nonresonance relations*

$$s_1\lambda_1(a) + s_2\lambda_2(a) + \ldots + s_N\lambda_N(a) \not\equiv 0 \tag{16}$$

hold for all $s \in Z^N$ such that $1 \le |s| \le M$ and $|s_{n+1}| + \cdots + |s_N| \le 2$.

Then for arbitrary $\rho < 1$ and for positive ε small enough there exist a Borel subset $\mathfrak{A}_\varepsilon(I) \subset \mathfrak{A}$ and analytic embeddings

$$\Sigma^\varepsilon_{a,I} : \mathrm{T}^n \longrightarrow Z \,, \quad a \in \mathfrak{A}_\varepsilon(I) \,, \tag{17}$$

such that

a) $\mathrm{mes}(\mathfrak{A} \backslash \mathfrak{A}_\varepsilon(I)) / \mathrm{mes}\,\mathfrak{A} \to 0$ $(\varepsilon \to 0)$;

b) the map $(q, I, a) \longmapsto \sum^\varepsilon_{a,I}(q)$ is Lipschitz and is ε^ρ-close to the map $(q, I, a) \longmapsto \Sigma^0(q, I)$;

c) for $a \in \mathfrak{A}_\varepsilon(I)$ the torus $\Sigma^\varepsilon_{a,I}(\mathrm{T}^n)$ is invariant for the equation (10) and is filled with quasiperiodic solutions of the form $u_\varepsilon(t) = \Sigma^\varepsilon_{a,I}(q + \omega_\varepsilon t)$ with a frequency vector $\omega_\varepsilon \in \mathrm{R}^n$ which is $C\varepsilon$-close to $\omega = (\lambda_1, \ldots, \lambda_n)$. All Lyapunov exponents of these solutions are equal to zero.

Refinement (see Part 3, Theorem 1.1). In the variables (13) the unperturbed hamiltonian is equal to $\omega \cdot p + \frac{1}{2}\langle A^Y y, y \rangle$ and the perturbation is $\varepsilon H_a(q, p, y)$. The statements of the theorem remain true for perturbations of the more general form

$$\varepsilon H_a = \varepsilon H_{1a}(q, p, y) + H^3_a(q, p, y) \,, \quad H^3_a = O\big(|p - I|^2 + \|y\|^3 + \|y\|\,|p - I|\big) \,.$$

This form of the result is suitable for applications to perturbations of nonlinear problems (see below).

The formulations of our results given above are "almost exact". For the exact statements see the main text. In Part 2 of the text we state local and global in a versions of Theorem 1 (Theorem 2.1.1 and 2.2.2 respectively, where the latter is a rather simple consequence of the former); we give various applications of the theorems to nonlinear perturbations of linear PDE's and postpone the proof of Theorem 2.1.1 till Part 3. There we reformulate the theorem in a more general form to facilitate its applications to nonlinear problems (the reformulated theorem also includes Refinement given above) and prove the result. In Part 3 we also give a version of Theorem 1, applicable to the problems with the natural parameter $\omega = (\lambda_1, \ldots, \lambda_n)$ varying in a domain of small diameter of order ε^κ, $0 < \kappa < 1$. This result is useful to study small-amplitude oscillations in nonlinear PDE's (see item 3.2.B below).

Remark. If the natural parameter ω is chosen for the parameter of the equation and λ_j does not depend on ω for $j \ge n + 1$, then the assumption 3) is fulfilled trivially. If in addition $\dim Z < \infty$, then the assumptions 1), 2) hold trivially, too. So for finite-dimensional systems (written in the form (14)) Theorem 1 coincides with Melnikov's theorem.

As another infinite-dimensional Melnikov-type theorem we mention the result of Wayne's paper [W1], devoted to the nonlinear string equation with a random potential. We discuss the approach, the work [W1] is based on, below.

Remark. If the hamiltonian of the perturbation is quadratic, then the theorem's statements for all parameters a (and ε small enough) immediately result from the classical perturbation theory for the discrete spectrum of a linear operator in Hilbert space (see, e.g. [RS]). In the nonlinear case the theorem's statements certainly does not hold for *all* parameters because of resonances between the frequencies $\{\lambda_j\}$, which occur for some a and give rise to much more complicated phenomenous. For discussions some of them in the finite-dimensional situation see [AKN] and [Mo].

Remark. As the map (17) is ε^ρ-close to the map $q \longmapsto \Sigma^0(q,I)$, then the solutions $u_\varepsilon(t)$ are ε^ρ-close to the curves $t \longmapsto \Sigma^0(q+\omega_\varepsilon t, I)$ for all t. The vector ω_ε is equal to $\omega+\varepsilon\omega_1+\varepsilon^2\omega_2+\cdots$, where the vector ω_1 may be obtained via some natural averaging (see [K4]). So Theorem 1 gives an averaging procedure for low-frequency solutions of equation (10) as a simple consequence.

Under the assumptions of the theorem an unperturbed torus $T^n(I)$ with

$$I \in \mathcal{I} = \{x \in \mathbb{R}^n \mid K^{-1} \le x_j \le K \ \forall j\}$$

survives in the equation (10) if $\varepsilon \le \varepsilon_0$ and a belongs to a set $\mathfrak{A}_\varepsilon(I)$ such that

$$\mathrm{mes}(\mathfrak{A}\backslash\mathfrak{A}_\varepsilon(I)) \le \nu(\varepsilon)\mathrm{mes}\,\mathfrak{A}\ ,$$

where $\nu(\varepsilon) \to 0$ as $\varepsilon \to 0$. The number ε_0 and the function $\nu(\varepsilon)$ do not depend on I (but depend on K). Let us denote

$$\mathcal{I}_\varepsilon(a) = \{I \in \mathcal{I} \mid a \in \mathfrak{A}_\varepsilon(I)\}\ .$$

The torus $T^n(I)$ persists if $I \in \mathcal{I}_\varepsilon(a)$. By Fubini theorem,

$$(\mathrm{mes}\,\mathfrak{A})^{-1}\int_{\mathfrak{A}} \mathrm{mes}(\mathcal{I}\backslash\mathcal{I}_\varepsilon(a))\,da = (\mathrm{mes}\,\mathfrak{A})^{-1}\int_{\mathcal{I}} \mathrm{mes}(\mathfrak{A}\backslash\mathfrak{A}_\varepsilon(I))\,dI \le \mathrm{mes}\,\mathcal{I}\,\nu(\varepsilon)\ . \tag{18}$$

Let us consider the sets

$$Z_K^0 = \{(q,I) \in Z^0 \mid I \in \mathcal{I}(a)\}\ , \quad Z_K^\varepsilon = \{(q,I) \in Z^0 \mid I \in \mathcal{I}_\varepsilon(a)\}\ .$$

By (18) for a typical a the relative measure of Z_K^ε in Z_K^0 is no less then $1-\nu(\varepsilon)$. The image of the set Z_K^ε under the map

$$(q,I) \longmapsto \Sigma_{a,I}^\varepsilon(q) \tag{19}$$

is invariant for the flow of equation (10) and is filled with quasiperiodic solutions. The mapping (19) is Lipschitz and ε^ρ-close to the embedding Σ^0. So the Hausdorf measure $\mathcal{H}^{2n}$ (see [Fe]) of the invariant set as above is no less then

$$(1-\nu_1(\varepsilon))\mathrm{mes}_{2n} Z_K^0\ , \tag{20}$$

with some $\nu_1(\varepsilon) \to 0$ as $\varepsilon \to 0$. Taking K large enough and ε sufficiently small one can make (20) as large as desired. So we have seen that under the assumptions of Theorem 1 for typical a and for ε small enough the equation (10) has invariant sets of the Hausdorf measure $\mathcal{H}^{2n}$ as large as desired. These sets are filled with quasiperiodic trajectories with zero Lyapunov exponents. They form obstacles to the fast stochastisation of solutions of a typical system of form (10). Our guess is that the recurrence effect "of FPU type" is caused by such sets.

Our results leave without any answer the natural question: do the *infinite-dimensional* invariant tori of the system (10) with $\varepsilon = 0$ persist under Hamiltonian perturbations? The answer is affirmative if the following three assumptions are satisfied:

a) the perturbation H has short range interactions, i.e. for $u(t)$ written as $\Sigma x_k^{\pm}(t)\varphi_k^{\pm}$, and for some finite N the equation for $x_k^{\pm}$ does not depend on $x_m^{\pm}$ with $|k-m| \geq N$ (or depends on $x_m^{\pm}$ in an exponentially small with respect to $|k-m|$ way);

b) $|H(u)| = O(\|u\|^d)$ for some $d > 2$;

c) the coefficients $x_k^{\pm}$ decrease, for example, exponentially when k is growing.

The assumptions a), b) never hold for nonlinear partial differential equations (but they are fulfilled for some equations from the physics of crystals). For the exact statements see [FSW], [VB] and [P2], [W2], [AlFS], [ChP]. We remark that the works [FSW], [VB] were the first ones where KAM theory was applied to infinite-dimensional Hamiltonian systems.

Without the assumptions a)-b) the maximal magnitude of the perturbation which allows one to prove Kolmogorov's theorem (=to prove preservation most of half-dimensional tori) exponentially decrease with the dimension of the phase-space (see e.g. [P2, p.364]). We suppose that the exponential estimate is the best possible one. In particular, infinite-dimensional tori "in general" do not survive under the system's perturbations.

We end this part with the remark that some results concerning the preservation of infinite-dimensional tori in equation (10) with the sectrum $\{\pm i\lambda_j\}$ of a special type may be obtained via infinite-dimensional versions of Siegel's theorem. See [War], [Z2] and especially [Nik].

3. Applications

In this item we show that Theorem 1 gives a flexible tool to study nonlinear Hamiltonian PDE's with one-dimensional spatial variable. We discuss applications to nonlinear perturbations of linear equations, to small-amplitude oscillations in nonlinear equations and to perturbations of the integrable PDE's. We end the item with some arguing why the theorem can not be applied to multi-dimensional PDE's.

3.1 Perturbations of linear differential equations

As a rule, the assumption 1) of Theorem 1 is fulfilled if JA is a differential operator on a segment with some self-adjoint boundary conditions. So the theorem is applicable to spatially one-dimensional quasilinear Hamiltonian partial differential equations, depending on a vector parameter.

Example 1 (see [K1] and Part 2.3). Let us consider nonlinear Schrödinger equation with a bounded real potential $V(x;a)$, depending on an n-dimensional parameter a:

$$\begin{aligned} &\dot{u} = i(-u_{xx} + V(x;a)u + \varepsilon\varphi'(x,|u|^2;a)u) , \\ &u = u(t,x) , \ t \in \mathbb{R} , \ x \in (0,\pi) ; \ \ u(t,0) \equiv u(t,\pi) \equiv 0 . \end{aligned} \tag{21}$$

Here φ is a real function analytic in $|u|^2$ and $\varphi' = \partial\varphi/\partial|u|^2$. To apply the theorem one has to set Z equal to the space of square-summable complex-valued functions on $(0,\pi)$ (and consider it as a real Hilbert space), to set A_a equal to the differential operator $-\partial^2/\partial x^2 + V(x;a)$ under the Dirichlet boundary condition, to set $Ju(x) = iu(x)$ and

$$H_a(u(x)) = \frac{1}{2}\int_0^\pi \varphi(x,|u(x)|^2;a)\,dx .$$

Let us denote by $\{\varphi_j(x;a)\}$, $\{\lambda_j(a)\}$ complete systems of real eigenfunctions and eigenvalues of the operator A_a. The invariant n-tori of the unperturbed problem are of the form

$$T(I) = \Big\{\sum_{j=1}^{n}(\alpha_j^+ + i\alpha_j^-)\varphi_j(x;a) \mid {\alpha_j^+}^2 + {\alpha_j^-}^2 = 2I_j > 0\ \forall j\Big\} .$$

By the well-known asymptotics of the spectrum of the Sturm–Liouville problem ([Ma], [PT]), $\lambda_j(a) = j^2 + O(1)$ and the assumption 1) of Theorem 1 is fulfilled with $d_1 = 2$, $\kappa = 3/2$. The theorem is applicable to the problem (21); therefore the torus $T(I)$ persists in the probelm (21) for most of a and ε small enough, if the potential V depends on a in a nondegenerate way. So for nondegenerate families of potentials $\{V(\cdot;a)\}$ and for typical parameters a equation (21) has a lot of quasiperiodic in t solutions, localized in the phase-space Z in a ε^ρ-neighborhood of the low-frequency tori $T(I)$.

Example 2 (see Part 2.4). We consider nonlinear Schrödinger equation with real random potential $V_\nu(x)$ under the Dirichlet boundary conditions:

$$\dot{u} = i(-u_{xx} + V_\nu(x)u + \varepsilon\varphi'(x,|u|^2)u) , \ \ u(t,-\pi) \equiv u(t,\pi) \equiv 0 , \tag{22}$$

where ν is a random parameter. We denote by $QP_\varepsilon = QP_\varepsilon(\nu) \subset Z$ the random subset of the phase space $Z = L_2(-\pi,\pi;\mathbb{C})$, equal to the union of all time quasiperiodic solutions with zero Lyapunov exponents (we treat the solutions as curves in Z). It occurs that if the potential V is x-periodic "with good randomness properties", then the set QP_ε is asymptotically dense in the phase space as $\varepsilon \to 0$: for any complex function $\mathfrak{z}(x)$

$$\operatorname{dist}(\mathfrak{z}(\cdot), QP_\varepsilon) \longrightarrow 0 \quad (\varepsilon \to 0)$$

in probability.

To prove this statement we treat (22) as an equation (21) with an infinite-dimensional parameter a. We 1) apply the results of Example 1 to construct invariant tori of dimensions 1,2,... ; 2) prove that the union of these finite-dimensional invariant tori is asymptotically dense in Z when $\varepsilon \to 0$.

This result explains (and predicts) long-time regular behaviour of "typical" solutions of (22), trapped by linearly-stable regular solutions from QP_ε. The solutions in QP_ε can be eternally approximated with accuracy $C\varepsilon$ by the quasiperiodic solutions of linear equation $(22)|_{\varepsilon=0}$ with the frequency vector ω replaced by some averaged vector ω_ε (see Corollary 2.1.1 in Part 2). We think that this result, which is also true for other hamiltonian PDE's with random coefficients, can be treated as a kind of averaging theorem for nonlinear PDE's.

Example 3 (see Part 2.5). Theorem 1 can be applied to study nonlinear perturbations of the quantized harmonic oscillator

$$\begin{aligned} \dot{u} &= i\Big(-u_{xx} + (x^2 + V_0(x;a))u + \varepsilon\nabla H_a(u)\Big)\,, \\ u &= u(t,x)\,,\ x \in \mathbb{R}\,,\ u(t,\cdot) \in L_2(\mathbb{R})\,, \end{aligned} \tag{23}$$

where the function V_0 vanishes at $x = \pm\infty$. The operator $A_a = -\partial^2/\partial x^2 + x^2 + V_0$ has a discrete spectrum $\{\lambda_j(a)\}$, which obeys Bohr's quantization law: $\lambda_j \sim C(j+1/2)$. Moreover, $|\lambda_j - C(j+1/2)| \le C_1 j^{-1/2}$. So the spectral asymptotic assumption holds with $d_1 = 1$ and Theorem 1 can be applied to (23), provided that the gradiental map $\nabla H_a(u)$ is of a negative order. In particular, if

$$H_a = \frac{1}{2}\int \varphi\big(|u * \xi(x)|^2; a\big)\, dx$$

($u * \xi$ is the convolution with a smooth real-valued function ξ, vanishing at infinity).

We can also consider perturbed unharmonic oscillator

$$\dot{u} = i\Big(-u_{xx} + (x^2 + \mu x^4 + V_0(x;a))u + \varepsilon\varphi'(x, |u|^2; a)u\Big)\,, \tag{24}$$

where $\mu > 0$. Now

$$\lambda_j = C_1\big(n + \frac{1}{2}\big)^{4/3} + C_2\big(n + \frac{1}{2}\big)^{2/3} + O(1)\,,$$

so the assumption 2) of Theorem 1 holds in a slightly generalized form with $d_1 = \kappa = 4/3$ (below in Part 3 the theorem is stated and proven with the spectral assumption exactly in this form). The first assumption of Theorem 1 holds with $d_H = 0$, $d_J = 0$, $d_1 = 4/3$. So typically equation (24) has many time-quasiperiodic solutions, if ε is small enough.

Example 4 (see [K2] and Part 2.6). Let us consider the equation of oscillations of a string with fixed ends in nonlinear-elastic media depending on n-dimensional parameter:

$$\begin{aligned} \ddot{w} &= \big(\partial^2/\partial x^2 - V(x;a)\big)w - \varepsilon\varphi_w(x, w; a)\,, \\ w &= w(t,x)\,,\quad 0 \le x \le \pi\,,\quad t \in \mathbb{R}\,;\quad w(t,0) \equiv w(t,\pi) \equiv 0\,. \end{aligned} \tag{25}$$

After some reduction (see Part 2.6) Theorem 1 is applicable to this problem with the choice $d_1 = 1$, $\kappa = 3/2$, $d_H = -1$. So in a nondegenerate case quasiperiodic in

t solutions of the unperturbed problem (25) with $\varepsilon = 0$ persist in the problem (25) for most of a and for ε small enough.

Concrete examples of nondegenerate potentials are given in Part 2.6 (in particular if $n = 1$, then one can take $V(x;a) = a$).

If $n = 1$, then the theorem deals with time-periodic solutions

$$w(t,x) = I\varphi_j(x;a)\sin\left(\sqrt{\lambda_j(a)}\,(t+q)\right)$$

of linear string equation $(25)|_{\varepsilon=0}$, depending on a one-dimensional parameter a (as above, $\{\varphi_j(x;a)\}$ and $\{\lambda_j(a)\}$ are eigenfunctions and eigenvalues of the operator $-\partial^2/\partial x^2 + V(x;a)$). These solutions persist in the perturbed equation (25) for most of a, if $\lambda_j'(a) \not\equiv 0$ and

$$m \cdot \sqrt{\lambda_j(a)} \not\equiv \sqrt{\lambda_N(a)}\,, \quad m \cdot \sqrt{\lambda_j(a)} \not\equiv \sqrt{\lambda_N(a)} \pm \sqrt{\lambda_M(a)}\,,$$

where m is an arbitrary integer and the numbers j, N, M are pairwise different. This statement is an infinite-dimensional analog of the Lyapunov theorem (see item 1.1 above) with an additional second-order nonresonance condition. Recently Craig and Wayne [CW] proved that the extra condition may be omitted provided that the functions V and φ are analytic in x.

Time-periodic solutions of nonlinear string equation under the Dirichlet boundary conditions have been studied by many authors (see the survey [Brel]). Under different restrictions on the nonlinear term of the equation it was proven that the equation has a *countable* family of time-periodic solutions. Our tools enable us to prove that for typical potentials the equation has time-periodic solutions, parametrized by the points of some *one-dimensional* sets (see (18)). In [BoK] similar result is proven for parameter-independent equation (25) with $V = 1$, provided that $\varphi = \kappa\varphi^4 + O(|\varphi|^5)$, $\kappa \neq O$ (the proof is based on an application of Theorem 1 to perturbations of the Sine-Gordon and Sinh-Gordon equations).

3.2 Perturbations of nonlinear systems

A) ***Perturbations of Birkhoff-integrable systems*** **(see [K5], Example 1).**

We call a Hamiltonian system *Birkhoff integrable* if it may be analytically reduced to an infinite sequence of Hamiltonian equations of the form

$$\dot{x}_j^+ = \partial H_0/\partial x_j^-\,, \quad \dot{x}_j^- = -\partial H_0/\partial x_j^+\,, \quad j = 1,2,\ldots$$

with

$$H_0 = H_0(p_1, p_2, \ldots)\,, \quad p_j = \frac{1}{2}(x_j^{+^2} + x_j^{-^2})$$

(i.e., it may be analytically reduced to the Birkhoff normal form, see [Mo], [SM]). The n-tori

$$T(p) = \{x | x_j^{+^2} + x_j^{-^2} = 2p_j\,, \ j = 1,\ldots,n\,; \ 0 = x_{n+1}^{\pm} = x_{n+2}^{\pm} = \cdots\}$$

are invariant for the system. It is convenient to pass to the variables (q,p,y) as in (13) with $p=(p_1,\dots,p_n)$, $y=(y_1^+,y_1^-,y_2^+,\dots)$, $y_j^\pm = x_{n+j}^\pm$ $(j=1,2,\dots)$. In these variables the equations have the form (14) with

$$\mathcal{H}_0(q,p,y) = h(p) + \frac{1}{2}\langle A(p)y,y\rangle + O(|y|^3)\,,$$

where

$$h(p) = H_0(p_1,\dots,p_n,0,\dots)\,,$$

and

$$\langle A(p)y,y\rangle = \sum_{j=1}^{\infty}(y_j^{+^2}+y_j^{-^2})\frac{\partial}{\partial p_{n+j}}h(p_1,\dots,p_n,0,\dots)\,.$$

So the ε-perturbed hamiltonian in the new variables is equal to

$$\mathcal{H}_\varepsilon = \mathcal{H}_0 + \varepsilon H_1 = h(p) + \frac{1}{2}\langle A(p)y,y\rangle + O(\|y\|^3) + \varepsilon H_1\,. \tag{26}$$

Let us fix for a moment some $a\in\mathbb{R}_+^n$ and rewrite $\mathcal{H}_\varepsilon$ as follows:

$$\mathcal{H}_\varepsilon = \left[h(a)-\omega(a)\cdot a\right]+\omega(a)\cdot p+\frac{1}{2}\langle A(a)y,y\rangle+\varepsilon H_1+O(\|y\|^3+|p-a|^2+|p-a|\,\|y\|^2)\,,$$

where $\omega(a)=\nabla h(a)$. The term in the square brackets does not affect the dynamics and may be neglected. Let us suppose that the system possesses nondegenerate *amplitude-frequency modulation*:

$$\operatorname{Hess} h(a) = \det\{\partial\omega_j(a)/\partial a_k\}\not\equiv 0\,. \tag{27}$$

Then one can treat the vector a as a parameter of the problem and apply to the perturbed problem Theorem 1, taking into account Refinement. So if spectral asymptotics and nondegeneracy assumptions are fulfilled, then most of invariant tori $T(a)$ survive under perturbations.

The trick we have just discussed is well suited to study perturbations of finite-dimensional integrable systems but not perturbations of integrable partial differential equations of Hamiltonian form. The reason is that in the last case the transition to the Birkhoff coordinates (or to the action-angle ones) is not regular.[5] To handle the integrable PDE's one needs more sophisticated approach; see item C) below.

B) ***Use of the partial Birkhoff normal form***

One can treat the unperturbed linear Hamiltonian system (10) as a Birkhoff integrable system with the quadratic hamiltonian $\mathcal{H}_0 = h(p)+\frac{1}{2}\langle Ay,y\rangle$, where $\frac{1}{2}\langle Ay,y\rangle = \frac{1}{2}\Sigma_{j=n+1}^{\infty}\lambda_j(x_j^{+^2}+x_j^{-^2})$ and $h(p)=\lambda_1p_1+\dots+\lambda_np_n$, $\omega(p)=(\lambda_1,\dots,\lambda_n)$.

[5] At least, the smoothness or analyticity of the action-angle variables is not proven yet. See in [McT] *continuous* action-angle variables for the KdV equation; see [Kap] for the fact that the constructed in [McT] foliation of the phase-space to invariant tori — not the action-angle variables themselves! — is analytic in L_2-norm.

Now the condition (27) is broken and one can not use an amplitude-frequency modulation to avoid outer parameters a. Nevertheless sometimes one can extract the modulation from the perturbation. This trick was successfully used in a number of works, starting (as far as we know) with Arnold's paper [A3] devoted to Hamiltonian systems with proper degeneration (see also [AKN]); Pöschel [P1] used the trick in his investigations of lower-dimensional tori, Wayne [W1] used similar approach to prove the existence of quasiperiodic in time solutions of nonlinear string equation with a random potential. Now we turn to its discussion.

For the sake of symplicity we restrict ourselves to the perturbations of the form $H = H^3 + H^4$ with homogenous of order j functions H^j, $j = 3,4$. Let us pass to the variables (13). Then the perturbed hamiltonian is $\mathcal{H}_\varepsilon = \mathcal{H}_0 + \varepsilon H_1$ with

$$H_1 = H^0(q,p) + \langle H^1(q,p), y\rangle + \frac{1}{2}\langle H^2(q,p)y, y\rangle + O(\|y\|^3)\,.$$

Here H^1 is a vector in Y and H^2 is a selfadjoint operator. So

$$\mathcal{H}_\varepsilon = h(p) + \frac{1}{2}\langle Ay, y\rangle + \varepsilon\big[H^0(q,p) + \langle H^1(q,p), y\rangle + \frac{1}{2}\langle H^2(q,p)y, y\rangle + O(\|y\|^3)\big]\,.$$

It is known since Birkhoff that with the help of a formally-analytic symplectic change of variables $\mathcal{H}_\varepsilon$ may be put into a partial normal form as follows:

$$\mathcal{H}_\varepsilon = h^1(p) + \frac{1}{2}\langle A^1(p)y, y\rangle + \varepsilon^2 H_\Delta(q,p,y) + \varepsilon O(\|y\|^3)\,.^{6)} \tag{28}$$

Here $h^1(p) = h(p) + \varepsilon \bar{H}^0(p)$ (the bar means the averaging over $q \in \mathrm{T}^n$) and $A^1 = A + \varepsilon A_\Delta(p)$ with some operator $A_\Delta(p)$ constructed in terms of the operator $\bar{H}^2(p)$. The function (28) is of the same form as (26) and in general the assumption (27) is fulfilled for the function $h^1(p)$.

The natural parameter $\omega = \nabla h^1(p)$ varies now in a domain of a small diameter δ_a, $\delta_a \sim \varepsilon$ (the perturbation is much smaller — of order ε^2). So Theorem 1 can not be directly applied to the equation (28). To handle this class of problems we state in Part 3 of the book Theorem 3.1.2, devoted to the equation (10) written in the variables (13), with the set $\mathfrak{A}$ equal to the ball of a small radius δ_a. The theorem states that the assertions of Theorem 1 remain true if

$$\varepsilon H_a = \varepsilon^{1+\mu} H_{1a}(q,p,y) + \delta_a H_a^3(q,p,y)\,,$$

$$H_a^3 = O\big(|p-I|^2 + \|y\|^3 + \|y\|^2\,|p-I|\big)\,,$$

where $\mu > 0$ and $1 \geq \delta_a \geq C\varepsilon$. (For the exact statement of the result see Part 3.1.)

The application of this result with $\delta_a = \varepsilon$, $\mu = 1$ to the equation with the hamiltonian (28) proves persistence most of the invariant n-tori $T^n(p)$ of the initial

6) One can achieve this normal form by formal applying to H_ε the transformation S_0 from Part 3.2 (the "KAM-step" of our proof), taking for granted that all the involved series converge.

linear system, provided that the transformation to the partial normal form converges and the nondegeneracy assumption

$$\text{Hess}\, h_1(p) \not\equiv 0$$

holds together with the nonresonance one (the assumption 3) of Theorem 1, where $\lambda = \tilde{\lambda}$, $(\tilde{\lambda}_1, \dots, \tilde{\lambda}_n)(p) = \nabla h_1(p)$ and $\{\tilde{\lambda}_{n+1}(p), \tilde{\lambda}_{n+2}(p), \dots\}$ is the spectrum of the operator $A^1(p)$). Below we call these assumptions *nonlinear*, because they reflect nonlinear nature of the equation with the hamiltonian (28) (the frequencies $\{\tilde{\lambda}_j\}$ of oscillations depend on their amplitude-vector p).

The exact formulae, which can be constructed as in [P1], or, due to the last footnote, can be extracted from the proof of Theorem 1 (see below Part 3.2 with $m = 0$ and Part 3.8), show that the normal-form transformation is defined as a series with some regular numerators and with denominators of the form $D(s) = s_1\lambda_1 + \dots + s_N\lambda_N$. Here N is an arbitrary natural number $\geq n+1$ and

$$3 \leq |s_1| + \dots + |s_N| \leq 4\,, \quad s_N \neq 0\,, \quad |s_{n+1}| + \dots + |s_N| \leq 2\,. \tag{29}$$

So if

$$|D(s)| \geq C^{-1} \tag{30}$$

for all s as above, then the normal-form transformation converges.

The condition (30) is not very restrictive because it holds for typical sequences $\{\lambda_j\}$ satisfying assumption 2) of Theorem 1. Now we check this statement for $d_1 > 1$. To do it let us take some $N \geq n+1$. Then

$$\begin{aligned} |D(s)| &\geq |\lambda_{n+1}s_{n+1} + \dots + \lambda_N s_N| - |\lambda_1 s_1 + \dots + \lambda_n s_n| \\ &\geq |\lambda_N - \lambda_{N-1}| - 3\max\{|\lambda_j| \mid 1 \leq j \leq n\} \geq C_1 N^{d_1-1} - C_2\,. \end{aligned}$$

So (30) holds with $C = 1$ if N is greater than some N_0. Therefore the inequality (30) holds if $D(s) \neq 0$ for the finite set of resonance relations consisting of all admissible relations with $N \leq N_0$ (one can choose C^{-1} equal to $\min\{1, \min\{|D(s)| \mid N \leq N_0\}\}$).

Thus, one can guarantee the convergence of the normal-form transformation for a parameter-dependent system for most values of the parameter. Due to simplicity of the involved resonance relations, often it is sufficient to have a *one-dimensional* parameter to obtain the convergence for *any fixed* $n \geq 1$.

The scheme we have just explained is applicable to study parameter-depending perturbed equation (10), if we

1) take away a set $\mathfrak{A}_c$ of parameters $a \in \mathfrak{A}$, violating the estimate (30) for some s as in (29) (this set is small, if C is large enough) and transform the equation to the normal form (28);

2) check the nonlinear nondegeneracy and nonresonance assumptions for $\{\tilde{\lambda}_j(p)\}$ and apply Theorem 3.1.2, treating $\omega = \nabla h^1(p)$ as the parameter.

The advantage of this approach to prove persistence of invariant n-tori of the linear system $(10)|_{\varepsilon=0}$ is that the set $\mathfrak{A}_c$ of "bad" parameters a is now an I- and

ε-independent set, which is relatively under the control. The disadvantage is the necessity to check the nonlinear nondegeneracy and the nonresonance assumptions for the "averaged" spectrum $\{\tilde\lambda_j(p)\}$, in addition to the ones for the initial spectrum $\{\lambda_j(a)\}$ which we need to make the first step.

This approach is applicable to study nonlinear Schrödinger and nonlinear string equations we discussed in Examples 1, 4 above. It is not difficult to check that the nonlinear nondegeneracy and nonresonance assumptions hold e.g., if in (25) the nonlinear term $-\varepsilon\varphi_w$ is equal to $-\varepsilon w^3$. The existence of time-quasiperiodic solutions of the equation

$$\ddot w = w_{xx} - V(x)w - \varepsilon w^3\ , \quad w(t,0) \equiv w(t,\pi) \equiv 0\ , \tag{31}$$

with the potential $V(x)$ lying outside a small set of "bad" potentials was obtained in a similar way by C.E. Wayne. In his paper [W1] the set of all potentials is given some Gaussian measure and the set of "bad" potentials is constructed as its small-measure subset.

In fact, the infinite-dimensional parameter $V(x)$, used in [W1], is much excessive: the scheme given above is applicable to (31) with $V(x) \equiv m \in \mathbb{R}^+$. It allows to prove existence of time-quasiperiodic solutions for most "masses" $m > 0$.

Remark. If the quadratic hamiltonian $\mathcal{H}_0$ is perturbed by higher-order terms starting from fourth order, then $\mathcal{H}_\varepsilon(z) = \mathcal{H}_0(z) + H_4(z) + H_5(z) + \cdots$. To study small-amplitude solutions of the corresponding Hamiltonian equation one can rescale $z = \mu u$, $\mu << 1$, obtain for u the equation with the hamiltonian $\tilde{\mathcal{H}}_\varepsilon(u) = \mathcal{H}_0(u) + \mu^2 H_4(u) + \mu^3 H_5(u) + \cdots$ and proceed exactly as above (with $\varepsilon = \mu^2$).

If the perturbation includes cubic terms, then the rescaled hamiltonian $\tilde{\mathcal{H}}_\varepsilon(u)$ contains the term $\mu H_3(u)$ which does not contribute to the function $h^1(p)$. Now the perturbation in (28) is larger than Hess h^1 (the perturbation is of order μ and the Hessian — μ^2); so we can not use $\omega = \nabla h^1$ as a parameter to apply the theorem.

C) *On the integrable equations of mathematical physics*

One of the main achievements of mathematical physics during the last decades was the discovery of theta-integrable nonlinear partial differential equations (see e.g. [DEGN], [NMPZ]). Such equations are quasilinear Hamiltonian equations of the form (9). They possess invariant symplectic $2n$-dimensional manifolds $\mathcal{T}^{2n}$ such that the restriction of the system (9) on $\mathcal{T}^{2n}$ is integrable. So $\mathcal{T}^{2n}$ is symplectomorphic to $\mathbb{T}^n_q \times P_p$, $P \subset \mathbb{R}^n$, and in coordinates (q,p) the restriction of the system onto $\mathcal{T}^{2n}$ has the form

$$\dot q = \nabla h(p)\ , \quad \dot p = 0\ .$$

Therefore $\mathcal{T}^{2n}$ is foliated into invariant n-tori $T^n(p) = \{(q,p) | p = \text{const}\}$ filled with quasiperiodic solutions $u_0(t) = \big(q + t\nabla h(p), p\big)$. The question is if the tori $T^n(p)$ survive under Hamiltonian perturbations of the equation. To formulate the corresponding result we have to consider variational equations about the solutions $u_0(t)$:

$$\dot v = J\Big(\nabla\mathcal{K}\big(u_0(t)\big)\Big)_* v$$

and to suppose that these equations are reducible to constant coefficient linear equations by means of a quasi-periodic substitutions $v = B(t,p)V$ (B is linear operator in Z quasiperiodically depending on t). It is proved (see [K3], [K5], [K8]) that under the reducibility assumption the quasilinear equation (9) near the manifold T^{2n} may be written in the form (14) with

$$\mathcal{H} = h(p) + \frac{1}{2}\langle A(p)y, y\rangle + O(\|y\|^3) .$$

A perturbed equation under this reduction takes exactly the form (26). So as in item A) one can prove that in a nondegenerate situation most of the tori $T^n(p)$ persist under perturbations.

The integrable nonlinear PDE's, linearized about their time-quasiperiodic solutions, *are* reducible to constant-coefficient equations. So Theorem 1 is applicable to study their perturbations. For an exact realization of this scheme for a perturbed Korteweg–de Vries and Sine–Gordon equations see [K5] and [BiK], [BoK].

See [K7] for a more detailed discussion of this group of applications of Theorem 1.

D) *Some remarks on multidimensional problems*

The most restrictive for applications among the assumptions of Theorem 1 is the assumption 2) (spectral asymptotics). As we have seen, this assumption holds for the differential equations with x-variable in a finite segment (or in the whole real line if the potential of the equation grows at infinity fast enough). The assumption 2) may be somewhat weakened with the same proof being applicable (see Remark 7 in Part 1.2). However, to carry out the proof the "separation condition"

$$\inf_{j\neq k} |\lambda_j(a) - \lambda_k(a)| \geq \delta > 0 \tag{$*$}$$

must be fulfilled (possibly, under some additional restriction one could replace (*) by the somewhat weaker assumption

$$|\lambda_j(a) - \lambda_k(a)| \geq \delta \max(j,k)^{-m} \quad \forall j \neq k\,, \tag{$**$}$$

with some "not too large" positive m).

We do not know any example where these conditions hold for a differential equation with multidimensional x. Conversely, (*) and (**) do not hold if the quantization arguments can be applied to construct quasimodes of the equation ([Laz], [GW]). In particular, (**) does not hold for the spectrum of the Dirichlet problem for the Laplace operator in a bounded convex two-dimensional domain with an analytic boundary [Laz].

Thus the nonlinear hamiltonian PDE's with x-variable in a segment form the distinguished class of equations with regular behavior of typical small-amplitude solutions.

4. Remarks on averaging theorems

Above we have proposed as an explanation for the recurrence effect of the FPU-type in partial differential equations of Hamiltonian form the theorem on persistence most of quasiperiodic solutions under Hamiltonian perturbations. It is well understood however that the long-time regular behavior of solutions may be explained by means of averaging theorems as well. In a finite-dimensional situation *Nekhoroshev's theorem* (see [N], [BGG], [Lo], [P4]) suites this purpose very well. For infinite-dimensional systems with discrete spectrum versions of this result are known only for systems with short range interactions ([W3], [BFG]). We are rather sceptical that there exists a version of Nekhoroshev's theorem applicable to nearly-integrable nonlinear partial differential equation.

First-order averaging theorems of Krylov–Bogolyubov type hold for a wide class of finite-dimensional systems. In the infinite-dimensional situation similar results are proven for lower-frequency initial data only (but for multidimensional in x equations also, see [Kri], [K3], [K4]). It is an open question if a first order averaging theorem for solutions of nearly integrable PDE's can be proven without this restriction.

5. Remarks on nearly integrable symplectomorphisms

Instead of differential equations (9) one can consider a "discrete-time equation" in the same infinite-dimensional phase-space $(Z, \alpha = -\langle J^{-1}dz, dz\rangle)$, i.e., a symplectic map

$$S : Z \longrightarrow Z\,, \quad S^*\alpha = \alpha\,. \tag{32}$$

The same phenomenon of pathologically regular behavior trajectories of nearly integrable sustem (32) (=iterations of the map S) can be observed; and the same question whether this phenomenon can be explained by existence many of finite-dimensional invariant tori of the map S appears.

A discrete-time theory parallel to the one for continuous-time systems we have discussed, can be developed. Fortunately, this work should not be done anew because of the following

Interpolation theorem. In the extended phase-space

$$\Big(Z \times \big\{(x,y) \in \mathbb{R}^2 \mid 1 < x^2 + y^2 < 2\big\}\,,\ \alpha \oplus dx \wedge dy\Big)$$

one can find a nearly integrable analytic Hamiltonian vector field with a hamiltonian $H(z,x,y)$ such that its isoenergetic Poincaré map with respect to the manifold $\{y=0\} \cap \{H = \text{const}\}$ is conjugated with S. In particular, if S is closed to the linear symplectic map exp JA, then H is close to $\frac{1}{2}\langle Az, z\rangle + \pi\big(|x|^2 + |y|^2\big)$.

A possible reformulation of the result is that *the nearly integrable analytic symplectic map S is conjugated with time-one shift along trajectories of analytic 1-periodic time-dependent Hamiltonian vector-field, close to an autonomous integrable one.*

So S inherits invariant finite-dimensional tori of the interpolating nearly integrable vector-field.

As far as we see, the constructive proof of a finite-dimensional version of this result, given in [K9] (see also [KP]), is also applicable in the infinite-dimensional setting. We did not include into [K9] an infinite-dimensional interpolation theorem mostly because we are not aware of any concrete infinite-dimensional symplectomorphism S of physical interest.

6. Notations

The list of notations we use is given at the end of the book. As usual, we refer to formula (2.3) form Part 1 as (1.2.3), if we are outside Part 1; we refer to Chapter 3.2 of Part 3 as to §2, if we are inside Part 3.

Acknowledgements

I am grateful to V.I. Arnold, V.F. Lazutkin and M.I. Vishik for useful discussions during the writing of the papers on which this book is based. I am indebted to Jürgen Pöschel for helpful remarks and friendly discussions during the preparation of the book; to M.B. Sevrjuk and Gene Wayne for pointing out a lot of misprints and mistakes; to Jürgen Moser and Eduard Zehnder for encouragement. I am much obliged to my wife Julia for encouragement and for typing the original Russian version of *all* my papers.

The draft of this text was written while I was visiting the Max–Planck–Institut für Mathematik in Bonn and was originally printed as the preprint of the MPI [K6]. The text was essentially revised and extended during my staying at the Forschungsinstitut für Mathematik (ETH, Zürich). I am thankful to both institutes for their warm hospitality, and to Rahel Boller (ETH) for her quality typing the text and her patience in carrying out countless corrections and improvements.

Part 1

Symplectic Structures and Hamiltonian Systems in Scales of Hilbert Spaces

The following notations are used below and everywhere in the book: for a Banach space Z the norm and the distance are denoted $|\cdot|_Z$ and dist_Z; if the space Z is Hilbert, the inner-product in Z is denoted $\langle\cdot,\cdot\rangle_Z$ (or just $\langle\cdot,\cdot\rangle$ if the space Z is fixed for a moment). For subdomains O_1, O_2 of Banach spaces the space of k-time Fréchet differentiable mappings from O_1 to O_2 is denoted $C^k(O_1; O_2)$. The tangent (cotangent) spaces for O_1, O_2 are identified with the corresponding Banach spaces (their conjugates spaces). For a map $\varphi \in C^k(O_1; O_2)$ the tangent (cotangent) map is denoted φ_* (φ^*). Vector-fields on O_1 are treated as the maps from O_1 to the corresponding Banach space. For a map $G: M_1 \to M_2$ of metric spaces M_1, M_2 we denote by $\text{Lip}G = \text{Lip}(G: M_1 \to M_2)$ its Lipschitz constant,

$$\text{Lip}\, G = \sup_{x\neq y} \frac{\text{dist}_2\big(G(x), G(y)\big)}{\text{dist}_1(x,y)}\,.$$

The complete list of the notations is given at the end of the book.

1. Symplectic Hilbert scales and Hamiltonian equations

The book is devoted to quasilinear Hamiltonian PDE's. The well-known dynamical system approach to quasilinear evolutionary PDE's (in particular, to the Hamiltonian ones) is to treat them as ordinary differential equations in the L_2-space of x-dependent functions; see e.g., [Lio]. In such a case usually the domain of definition of the corresponding vector-field in the L_2-space is some Sobolev function space (in particular, the space of functions with the derivatives squared-integrable up to some order). To study properties of the solutions (in addition to their existence), one should use in the researches some additional function spaces.

The natural formalisation of this approach is to consider Hamiltonian equations in a Hilbert space Z with given system of dense Hilbert subspaces Z_s, $s \geq 0$, such that $Z_s \subset Z_r$ if $s > r$ and $Z_0 = Z$. The norm, the distance and the scalar product in Z_s are denoted $\|\cdot\|_s$, dist_s and $\langle\cdot,\cdot\rangle_s$ respectively (in particular, $\|\cdot\|_0 = \|\cdot\|_Z$, $\text{dist}_0 = \text{dist}_Z$ and $\langle\cdot,\cdot\rangle_0 = \langle\cdot,\cdot\rangle_Z$). A vector $z \in Z$ defines the linear functional $\langle z,\cdot\rangle_Z$ in the space Z_s. We denote by $\|z\|_{-s}$ its norm,

$$\|z\|_{-s} = \sup\{\langle z, z'\rangle \mid \|z'\|_s \leq 1\}\,,$$

and denote by Z_{-s} the completion of Z in this norm. The space Z_{-s} is conjugate to Z_s. So it is isomorphic to Z_s (via the Riesz's isomorphism) and carries the natural structure of a Hilbert space. We denote by $\langle\cdot,\cdot\rangle_{-s}$ the inner product in Z_{-s} and denote by $\langle\cdot,\cdot\rangle_Z$ (or $\langle\cdot,\cdot\rangle$ for short) the pairing of the spaces Z_s and Z_{-s}.

The set of Hilbert spaces $\{Z_s | s \in \mathbb{R}\}$ is called a scale of Hilbert space, or just a Hilbert scale, see [RS, LM].

We introduce the limit linear spaces Z_∞ and $Z_{-\infty}$,

$$Z_\infty = \cap Z_s\,, \quad Z_{-\infty} = \cup Z_s\,,$$

and give them no topology. We suppose that Z_∞ is dense in each space Z_s.

Definition. A system of continuous linear maps

$$L_s : Z_s \longrightarrow Z_{s-d}\,, \quad s \in \mathbb{R}\,, \tag{1.1}$$

is called a morphism of the Hilbert scale $\{Z_s\}$ of order d, if $L_{s_1} = L_{s_2}$ on $Z_{s_1} \cap Z_{s_2}$ for all s_1, s_2. Morphism (1.1) is called an isomorphism of order d, if each map L_s is an isomorphism.

For short, in what follows we omit index s in the morphism's notations.

Example 1.1. Let Z be the L_2-space of real 2π-periodic functions $u(x)$, $x \in \mathbb{T}^1 = \mathbb{R}/2\pi\mathbb{Z}$,

$$u(x) = \sum_{k\in\mathbb{Z}} e^{ikx} u_k\,, \; u_k = \bar{u}_{-k}\,, \; \|u\|_Z^2 = 2\pi \sum |u_k|^2 \;;$$

and let Z_s, $s \geq 0$, be the Sobolev space $H^s(\mathbb{T}^1)$ of functions with finite norm $\|\cdot\|_s$,

$$\|u\|_s^2 = 2\pi \sum |u_k|^2 (1 + k^2)^s$$

(so for $s \in \mathbb{N}$ Z_s is the space of functions with squared-integrable derivatives up to the order s). Norms $\|\cdot\|_{-s}$ with negative s are defined by the same formula. A space Z_∞ is now equal to the space of smooth periodic functions. For this example see also [Ch-B, RS].

To consider Hamiltonian equations, we should provide the Hilbert scale $\{Z_s\}$ with a symplectic structure. To do it let us suppose that we are given a linear operator $J : Z_\infty \to Z_\infty$ such that $J(Z_\infty) = Z_\infty$ and

a) J determines an isomorphism of the scale $\{Z_s\}$ of order $d_J \geq 0$;

b) the operator J with the domain of definition Z_∞ is antisymmetric in Z, i.e., $\langle Jz_1, z_2\rangle_Z = -\langle z_1, Jz_2\rangle_Z \;\; \forall z_1, z_2 \in Z_\infty$.

Let us denote by $\bar{J}$ the isomorphism of the scale $\{Z_s\}$ of order $-d_J$, equal to $-(J)^{-1}$.

Lemma 1.1. The operator $\bar{J} : Z \to Z_{d_J} \subset Z$ is bounded and antiselfadjoint in Z.

Proof. The first statement is evident because the space Z_{d_J} is continuously imbedded into Z. Let $x, y \in Z_\infty$ and $Jx = x_1$, $Jy = y_1$. Then $\bar{J}x_1 = -x$, $\bar{J}y_1 = -y$ and

$$\langle x_1, \bar{J}y_1\rangle = -\langle Jx, y\rangle = \langle x, Jy\rangle = -\langle \bar{J}x_1, y_1\rangle .$$

The operator $\bar{J} : Z \to Z$ is continuous, and the space Z_∞ is dense in Z, so $\bar{J}$ is antiselfadjoint and the lemma is proved. □

Let us introduce in every space Z_s with $s \geq 0$ the antisymmetric 2-form $\alpha = \langle \bar{J}\, dz, dz\rangle_Z$. By definition,

$$\langle \bar{J}\, dz, dz\rangle[z_1, z_2] = \langle \bar{J}\, z_1, z_2\rangle \;\forall z_1, z_2 \in Z_s\, . \tag{1.2}$$

Trivially, the form α is closed and nondegenerate [A, Ch-B, Laz].

Definition. The triple $\{Z, \{Z_s | s \in \mathrm{R}\}, \alpha = \langle \bar{J}\, dz, dz\rangle\}$ is called a symplectic Hilbert scale (or SHS for brevity).

The complexification $\{Z_s^c | s \in \mathrm{R}\}$ of the Hilbert scale $\{Z_s\}$ is defined in the natural way: $Z_s^c = Z_s \otimes_{\mathrm{R}} \mathrm{C}$. For all s we extend the scalar product $< \cdot, \cdot >_s$ to the complex-linear map $Z_s^c \times Z_s^c \to \mathrm{C}$, extend $\bar{J}$ to the complex-linear isomorphism of the complexified scale and extend α to the 2-form in Z_s^c, $s \geq 0$, by means of the formula (1.2).

Example 1.2. Let $Z = \mathrm{R}_p^n \times \mathrm{R}_q^n$, $Z_s = Z\ \ \forall s$ and $J : (p, q) \longmapsto (-q, p)$. In this case $J^2 = -E$, so $\bar{J} = -J^{-1} = J$, $d_J = 0$ and

$$\alpha = \langle \bar{J}\, dz, dz\rangle = \langle J\, dz, dz\rangle = dp \wedge dq\, .$$

Properties a), b) are obvious and we obtain the classical symplectic structure for even dimensional spaces [A, Laz].

Example 1.3. Let $Z = L_2(\mathrm{T}^1) \times L_2(\mathrm{T}^1)$ be the space of pairs of square-summable 2π-periodic functions $(p(x), q(x))$ and $Z_s = H^s(\mathrm{T}^1) \times H^s(\mathrm{T}^1)$ (see Example 1.1).

Let us take for a linear isomorphism J the operator

$$J : Z_s \longrightarrow Z_s\, ,\ (p(x), q(x)) \longmapsto (-q(x), p(x))\, .$$

Then $\bar{J} = J$ is an isomorphism of the scale $\{Z_s\}$ of order zero. Properties a), b) are evident.

Example 1.4. Let us set

$$Z_s = \left\{ u(x) \in H^s(\mathrm{T}^1) \mid \int_0^{2\pi} u(x) dx = 0 \right\}$$

for $s \in \mathrm{R}$, and $J = \partial/\partial x$. Then J is an isomorphism of the scale of order one and $\bar{J} = -(J)^{-1}$ is an isomorphism of order -1. Properties a), b) are evident again and we get SHS, corresponding to the symplectic structure of Korteweg–de Vries (KdV) equation (see below and [A, Appendix 13; NMPZ]).

For a domain $O_s \subset Z_s$ and a function $f \in C^1(O_s)$ let $\nabla f \in Z_{-s}$ be the gradient of f with respect to the inner product in Z:

$$\langle \nabla f(u), v\rangle = f_*(u)(v) = \frac{\partial}{\partial \varepsilon} f(u+\varepsilon v)\,\Big|_{\varepsilon=0}\ \forall v \in O_s\,.$$

The gradiental mapping $O_s \longrightarrow Z_{-s}$, $u \longmapsto \nabla f(u)$, is continuous.

Definition (cf. [A],[Ch-B],[ChM]). For $H \in C^1(O_s)$, the Hamiltonian vector-field V_H with the hamiltonian h is the map $V_H : O_s \longrightarrow Z_{-\infty}$ such that

$$\alpha\big(\xi, V_H(u)\big) = \langle \xi, \nabla H(u)\rangle \quad \forall \xi \in Z_\infty\,.$$

By the definition of V_H,

$$\langle J\xi, V_H(u)\rangle = \langle \xi, \nabla H(u)\rangle \quad \forall \xi \in Z_\infty\,.$$

So

$$V_H(u) = J\nabla H(u)\,,$$

and the Hamiltonian equation with hamiltonian H has the form

$$\dot u = J\nabla H(u)\,. \tag{1.3}$$

Let us denote by $D_s(V_H)$ the domain of definition of V_H in O_s

$$D_s(V_H) = \big\{u \in O_s \,|\, V_H(u) \in Z_s\big\}\,.$$

Definition (cf. [Bre]). A curve $u(t)$, $0 \le t \le T$, is called a strong solution of the equation (1.3) in the space Z_s iff $u \in C^1\big([0,T]; Z_s\big)$, $u(t) \in D_s(V_H)$ for all t and the equation (1.3) is satisfied. A curve $u \in C\big([0,T]; Z_s\big)$ is called a weak solution of (1.3) iff it is the limit in the $C\big([0,T]; Z_s\big)$-norm of some sequence of strong solutions.

Definition. Let O_s^1 be a subdomain of O_s such that for every $u_0 \in O_s^1$ there exists the unique weak solution $u(t) = S^t(u_0)$ $(0 \le t \le T)$ of equation (1.3) with initial condition $u(0) = u_0$. The set of the mappings

$$S^t : O_s^1 \longrightarrow O_s\,,\ u_0 \longmapsto S^t(u_0) \quad (0 \le t \le T)\,,$$

is called the "local semiflow of equation (1.3)" or the "flow of equation (1.3)" for short.

Weak solutions of equations (1.3) are generalized solutions in the sense of distributions (see [Lio] for a systematic use of this type of solutions):

Proposition 1.4. Let us suppose that for some $s_1 \in \mathbb{R}$ the map $J\nabla H : O_s \to Z_{s_1}$ is Lipschitz. Then a weak solution $u(t) \in O_s$ $(0 \le t \le T)$ of equation (1.3) is the generalized solution and after substitution of $u(t)$ into (1.3) the left and the right hand sides of the equation coincide as elements of the space $D'\big((0,T);Z_{s_2}\big)$ of distributions on $(0,T)$ with values in Z_{s_2}, $s_2=\min\{s,s_1\}$.

Proof. By the definition of a weak solution there exists a sequence of strong solutions $u_n(t)$ converging to $u(\cdot)$ in $C\big([0,T];X_s\big)$. For this sequence $\dot{u}_n\to\dot{u}$ in $D'\big((0,T);Z_s\big)$ and $J\nabla H(u_n)\to J\nabla H(u)$ in $C\big([0,T];Z_{s_1}\big)$. After transition to the limit in equation (1.3) one obtains the result. □

Example 1.2, again. Let $H\in C^1\big(\mathbb{R}_p^n\times\mathbb{R}_q^n\big)$. The Hamiltonian equation takes the classical form:

$$\dot{p}=-\nabla_q H(p,q)\,,\ \dot{q}=\nabla_p H(p,q)\,.$$

If $H\in C^2(\mathbb{R}^{2n})$, then a weak solution is the strong one and it exists for $0\le t\le T$, where T depends on the initial data $\big(p(0),q(0)\big)$.

Example 1.3, again. Let us consider the hamiltonian

$$H=\frac{1}{2}\int_0^{2\pi}\Big(p_x(x)^2+q_x(x)^2+V(x)\big(p(x)^2+q(x)^2\big)+\chi\big(p(x)^2+q(x)^2\big)\Big)\,dx$$

with smooth functions V and χ. Then $H\in C^1(Z_s)$ for $s\ge1$ and

$$\nabla H(p,q)=\big(-p_{xx}+V(x)p+\chi'(p^2+q^2)p\ ,\ -q_{xx}+V(x)q(x)+\chi'(p^2+q^2)q\big)\,.$$

The equation (1.3) takes now the following form:

$$\dot{p}=q_{xx}-V(x)q-\chi'(p^2+q^2)q\ ,\ \dot{q}=-p_{xx}+V(x)p+\chi'(p^2+q^2)p\,.$$

Let us denote $u(t,x)=p(t,x)+iq(t,x)$. The last equations are equivalent to the nonlinear Schrödinger equation with the real potential $V(x)$ for the complex function $u(t,x)$:

$$\dot{u}=i\big(-u_{xx}+V(x)u+\chi'(|u(x)|^2)u\big)\ ,\ u(t,x)\equiv u(t,x+2\pi)\,. \tag{1.4}$$

The equation (1.4) has the unique strong solution $u(t,x)$, $u(t,\cdot)\in Z_s$, $0\le t\le T=T\big(u(0,x)\big)$, if $s\ge1$ and $u(0,x)\in Z_{s+2}$ (we interpret here Z_s as a Sobolev space of periodic complex-valued functions), and (1.4) has the unique weak solution for $0\le t\le T$ if $u(0,x)\in Z_s$. For a simple proof see part 3 below.

Example 1.4, again. In the situation of example 1.4 let us consider the hamiltonian

$$H=\int_0^{2\pi}\big(\frac{1}{2}u_x^2+u^3\big)\,dx\,.$$

Then $H\in C^1(Z_s)$ provided that $s\ge1$, and

$$\nabla H\big(u(x)\big)=-u_{xx}+3u^2\,.$$

So now equation (1.3) is the KdV equation

$$\dot{u}(t,x) = -u_{xxx} + 6uu_x \tag{1.5}$$

for periodic in x functions with zero mean value:

$$u(t,x) \equiv (t, x+2\pi), \quad \int_0^{2\pi} u(t,x)\,dx \equiv 0 . \tag{1.5'}$$

It is well known [Ka], that for $s \geq 3$ the problem (1.5), (1.5′) has the unique strong solution $u(t,x)$, $u(t,\cdot) \in Z_s$, for every initial data $u(0,x) = u_0(x) \in Z_{s+3}$, and has the unique weak solution for every $u_0(x) \in Z_s$. The flow of problem (1.5), (1.5′) defines homeomorphisms of the phase space Z_s $(s \geq 3)$.[7)]

It is worth mentioning that any Hamiltonian equation (including (1.4) and (1.5), (1.5′)) may be written down in the form (1.3) in many different ways. For this statement see below Corollary 2.3.

2. Canonical transformations

Let $\{X, \{X_s\}, \alpha^X\}$ and $\{Y, \{Y_s\}, \alpha^Y\}$ be two SHS with 2-forms $\alpha^X = \langle \bar{J}^X dx, dx\rangle_X$ and $\alpha^Y = \langle \bar{J}^Y dy, dy\rangle_Y$ respectively; let $\bar{J}^X$ and $\bar{J}^Y$ be isomorphisms of the scales $\{X_s\}$ and $\{Y_s\}$ of order $-d_X$ and $-d_Y$ with $d_X, d_Y \geq 0$. Let $\varphi : O_X \to O_Y$ be a C^1-diffeomorphism of domains $O_X \subset X_{s_X}$ and $O_Y \subset Y_{s_Y}$, where $s_X, s_Y \geq 0$.

Definition. The diffeomorphism φ is called a canonical (or symplectic) transformation iff it transforms the 2-form α^Y into the 2-form α^X:

$$\phi^* \alpha^Y = \alpha^X . \tag{2.1}$$

Proposition 2.1. A C^1-diffeomorphism $\phi : O_X \to O_Y$ is canonical iff

$$\phi^* \bar{J}^Y \phi_* \equiv \bar{J}^X \tag{2.2}$$

(the identity holds in the space of linear operators $L(X_{s_X}; X_{-s_X})$).

Proof. From (2.1) one has for $v \in O_X$ and $\xi_1, \xi_2 \in X_{s_X}$ the identity

$$\langle \bar{J}^Y \phi_*(v)\xi_1, \phi_*(v)\xi_2\rangle_Y = \langle \bar{J}^X \xi_1, \xi_2\rangle_X . \tag{2.3}$$

Therefore

$$\langle \phi^*(v)\bar{J}^Y \phi_*(v)\xi_1, \xi_2\rangle_X = \langle \bar{J}^X \xi_1, \xi_2\rangle_X$$

for all $\xi_1, \xi_2 \in X_{s_X}$, and we have obtained (2.2). The backward implication from (2.2) to (2.1) can be obtained similarly. □

For a linear isomorphism $L : X_a \to Y_b$ a simple and useful criterion of the symplecticity can be stated in terms of symplectic basises. We remind that a system

---

7) In fact, the flow of the KdV equation (1.5), (1.5′) defines homeomorphisms of the L_2-space Z_0 [Bour], so the weak solutions are well-defined for u_0 in Z_0.

of vectors $\{\psi_j^\pm | j \in \mathbb{N}\}$ in X_a is called a basis if each vector in X_a can be uniquely represented by the convergent in X_a series $\psi = \Sigma x_j^\pm \psi_j^\pm$ with some numbers $x_j^\pm$.

Definition. A basis $\{\psi_j^\pm | j \in \mathbb{N}\}$ of X_a is called symplectic, if for $j,k \in \mathbb{N}$ and $\alpha, \beta = \pm$ the skew product $\alpha^X[\psi_j^\alpha, \psi_k^\beta]$ vanish, unless $j = k$ and $\alpha = -\beta$. In the latter case the skew product is equal to -1 (if $\alpha = +, \beta = -$) or 1 (if $\alpha = -, \beta = +$).

Proposition 2.1′. Linear isomorphism L is canonical iff it transforms a symplectic basis $\{\psi_j^\mp\}$ of the space X_a into the symplectic basis of the space Y_b.

We omit a trivial proof.

As in the finite-dimensional situation [A, Laz], a canonical transformation transforms solutions of a Hamiltonian equation into the solutions of the equation with the transformed hamiltonian:

Theorem 2.2. Let $\phi : O_X \to O_Y$ be a canonical transformation and let a curve $y : [0,T] \to O_Y$ be a strong solution of a Hamiltonian equation

$$y = V_{H^Y}(y) = J^Y \nabla H^Y(y)\,, \quad H^Y \in C^1(O_Y; \mathbb{R})\,. \tag{2.4}$$

Then $x(t) = \phi^{-1}(y(t))$ is the strong solution in O_X of the Hamiltonian equation

$$\dot{x} = V_{H^X}(x) = J^X \nabla H^X(x)\,, \tag{2.5}$$

with the transformed Hamiltonian $H^X = H^Y \circ \phi$. If the inverse mapping $\phi^{-1} : O_Y \to O_X$ is Lipschitz and y is a weak solution of (2.4), then x is the weak solution of (2.5).

Proof. As the gradient of the transformed hamiltonian H^X is equal to $\nabla H^X = \phi^* \nabla H^Y$ and the curve $x : [0,T] \to O_X$ is C^1-smooth, then the substitution into (2.4) $y(t) = \phi(x(t))$ gives the equation on $x(t)$:

$$\phi_* \dot{x} = \dot{y} = J^Y \nabla H^Y(y) = J^Y (\phi^*)^{-1} \nabla H^X(x)\,,$$

or

$$\dot{x} = (\phi_*)^{-1} J^Y (\phi^*)^{-1} \nabla H^X(x)\,,$$

where the right-hand side is well defined because

$$J^Y (\phi^*)^{-1} \nabla H^X(x) \in C([0,T]; O_Y)\,.$$

By (2.2), $J^X = (\phi_*)^{-1} J^Y (\phi^*)^{-1}$. Thus, $\dot{x} = J^X \nabla H^X(x)$, as stated.

The second statement of the theorem follows from the first one and the definition of a weak solution, because the mapping ϕ^{-1} is Lipshitz. □

Let $\{Y, \{Y_s\}, \alpha^Y\}$ be a SHS, let L be an isomorphism of the scale $\{Y_s\}$ of order $\Delta \le \frac{1}{2} d_Y$. Let us define the second SHS $\{X, \{X_s\}, \alpha^X\}$, where $X = Y$, $X_s = Y_s$ and $\alpha^X = \langle J^X dx, dx \rangle_X$, $J^X = L^* J^Y L$. The operator $\bar{J}^X$ is antisymmetric in X and it defines an isomorphism of the scale $\{X_s\}$ of the order $-d_Y + 2\Delta \le 0$, so the new triple is a SHS indeed. Let O_X be a domain in X_{s_X} and $O_Y \subset Y_{s_Y}$ be its image under the map L, where $s_Y = s_X - \Delta$. The mapping $L : O_X \to O_Y$ is canonical due to Proposition 2.1. So we have a trivial

Corollary 2.3 (change of the symplectic structure). Let $H^Y \in C^1(O_Y)$ and let $y(t) \in O_Y$ $(0 \le t \le T)$ be a solution of equation (2.4) (strong or weak). Then the curve $x(t) = L^{-1}y(t)$ is the solution of the Hamiltonian equation

$$\dot{x} = J^X \nabla H^X(x) ,$$

with the transformed hamiltonian $H^X = H^Y \circ L \in C^1(O_X)$, where $J^X = L^{-1}J^Y(L^*)^{-1}$.

Let $\{X, \{X_s\}, \alpha = \langle \bar{J}\, dx, dx\rangle_X\}$ be a SHS, $O^1 \subset O$ be domains in X_s and

$$\text{dist}_s(O^1; X_s \backslash O) \ge \delta > 0 . \tag{2.6}$$

Let $H \in C^2(O)$ and

$$\nabla H \in C^1(O; X_{s+d_J}), \quad \|J\nabla H(x)\|_s \le K, \quad \text{Lip}(J\nabla H : O \longrightarrow X_s) \le K . \tag{2.7}$$

Let us consider the Hamiltonian equation with the Hamiltonian H:

$$\dot{x} = J\nabla H(x) . \tag{2.8}$$

The estimates (2.6), (2.7) jointly imply that for $0 \le t \le T = \delta/K$ the flow of equation (2.8) defines the maps $S^t \in C^1(O^1; O)$ which are C^1-diffeomorphisms onto their images.

Theorem 2.4. For every $0 \le t \le T$ the mapping S^t is a canonical transformation.

Proof. We should prove that

$$(S^t)^* \alpha(x)[\eta_1, \eta_2] = \alpha[\eta_1, \eta_2] \quad \text{for} \quad x \in O^1 \quad \text{and} \quad \eta_1, \eta_2 \in X_s .$$

Since $S^0 = Id$, it is sufficient to check that

$$(S^{\cdot})^* \alpha(x)[\eta_1, \eta_2] = \text{const} \tag{2.9}$$

(i.e., the time-shifted symplectic form does not depend on the time). Let $x(\tau)$ be the solution of equation (2.8) with $x(0) = x$, and $\eta^j(t)$ $(j = 1, 2)$ be the solution of the Cauchy problem for the linearized about $x(\cdot)$ equation:

$$\dot{\eta}^j(\tau) = J(\nabla H)_*(x(\tau))\eta^j(\tau) , \; \eta^j(0) = \eta_j .$$

Then $(S^\tau)_*(x)\eta_j = \eta^j(\tau)$, $j = 1, 2$ and

$$(S^\tau)^* \alpha(x)[\eta_1, \eta_2] = \alpha\left[\eta^1(\tau), \eta^2(\tau)\right] = \langle \bar{J}\eta^1(\tau), \eta^2(\tau)\rangle =: \ell(\tau) .$$

The function $\ell(\tau)$ is continuously differentable. So to check (2.9) we should verify that $d/d\tau\, \ell(\tau) \equiv 0$. We have:

$$\begin{aligned}\frac{d}{d\tau}\ell(\tau) &= \langle \bar{J}\dot{\eta}^1, \eta^2\rangle + \langle \bar{J}\eta^1, \dot{\eta}^2\rangle = \\ &= \langle \bar{J}J(\nabla H)_*(x)\eta^1, \eta^2\rangle + \langle \bar{J}\eta^1, J(\nabla H)_*(x)\eta^2\rangle = \\ &= -\langle (\nabla H)_*(x)\eta^1, \eta^2\rangle + \langle \eta^1, (\nabla H)_*(x)\eta^2\rangle = 0\,,\end{aligned}$$

because the operator $\bar{J}$ is antiselfadjoint (Lemma 1.1) and the operator $(\nabla H)_*$ is selfadjoint. The theorem is proved. □

Let H_1, H_2 be C^1-functionals on a domain $O \subset X_s$. The gradiental maps $\nabla H_1, \nabla H_2$ are continuous X_{-s}-valued. Let us suppose that $\nabla H_j \in C(O; X_{s_j})$, $j = 1, 2$, where $s_1, s_2 \geq -s$ and $s_1 + s_2 \geq d_J$.

Definition. The Poisson bracket of the functionals H_1, H_2 as above is the continuous on O functional $\{H_1, H_2\}$ equal to

$$\{H_1, H_2\} = \langle J\nabla H_1, \nabla H_2\rangle_X\,.$$

Let $0 < \varepsilon \leq 1$ and H be a C^2-functional on O, satisfying (2.7). Let O^1 be a subdomain of O satisfying (2.6) and $S^t \in C^1(O^1; O)$, $0 \leq t \leq T = \delta/K$, be the flow of the equation

$$\dot{x} = \varepsilon J\nabla H(x)\,.$$

Theorem 2.5. If G is a C^1-functional on O, then

$$G\big(S^{t+\tau}(x)\big) = G(S^t x) + \tau\varepsilon\{H, G\}\big(S^t(x)\big) + O(\varepsilon\tau)^2$$

for $x \in O^1$ and $0 \leq t \leq t + \tau \leq T$. In particular, $\frac{d}{dt}G(S^t x) = \varepsilon\{H, G\}(S^t x)$.

Proof. From the assumptions imposed on H it follows that for $x \in O^1$ and $y = S^t x$

$$S^\tau(y) = y + \tau\varepsilon J\nabla H(y) + O(\tau\varepsilon)^2 \quad \text{in} \quad X_s\,.$$

So

$$\begin{aligned}G\big(S^\tau(y)\big) - G(y) &= \langle \nabla G(y), S^\tau(y) - y\rangle + O\,\|S^\tau x - x\|_s^2 \\ &= \tau\varepsilon\langle \nabla G(y), J\nabla H(y)\rangle + O(\varepsilon\tau)^2\,,\end{aligned}$$

and the theorem is proved. □

3. Local solvability of Hamiltonian equations

Let $\{Y, \{Y_s\}, \alpha\}$ be SHS, let O be a domain in Y_s, $s \geq 0$, and let H be a C^2-functional on O of the form

$$H(y) = \frac{1}{2}\langle Ay, y\rangle_Y + H_0(y)\,.$$

Here A is an isomorphism of the scale $\{Y_s\}$ of order $d_A \geq 0$ and A defines a selfadjoint operator in Y with the domain of definition Y_{d_A}. So $\nabla\left(\frac{1}{2}\langle Ay, y\rangle\right)(y) = Ay$, and the Hamiltonian equation with the hamiltonian H has the form

$$\dot{y} = J\left(Ay + \nabla H_0(y)\right) . \tag{3.1}$$

We shall prove a simple theorem on the local solvability of the equation (3.1) which will suit well to our aims. To formulate the theorem let us suppose that

$$\mathrm{Lip}(J\nabla H_0 : O \longrightarrow Y_s) \leq K \tag{3.2}$$

and let us fix subdomains $O^2 \subset O^1$ of O such that

$$\mathrm{dist}_s(O^1, Y_s \backslash O) \geq \delta > 0 . \tag{3.3}$$

Theorem 3.1. Let us suppose that the operator JA is antisymmetric in Y_s. Suppose also that every strong solutions $y(t)$ of equation (3.1) with initial data $y(0) = y_0 \in O^2$ stays inside domain O^1 for $0 \leq t \leq T$. Then for $y_0 \in O^2 \cap Y_{s+d_1}$, where $d_1 = d_A + d_J$, there exists the unique strong solution $y(t)$, $0 \leq t \leq T$, and for $y_0 \in O^2$ there exists the unique weak solution $y(t)$, $0 \leq t \leq T$.

Proof. Let us extend the mapping $J\nabla H_0 : O^1 \longrightarrow Y_s$ to a Lipschitz map $V : Y_s \longrightarrow Y_s$. One may take, for example,

$$V(y) = \begin{cases} \chi(y) J\nabla H_0(y), & y \in O , \\ 0, y \notin O , \end{cases}$$

where $\chi(y) = \delta^{-1} \max\left(0, \delta - \mathrm{dist}_{Y_s}(y; O^1)\right)$ (see (3.3)). The function χ is Lipschitz, it is equal to one in O^1 and to zero out of O. So the map V is Lipschitz and is equal to $J\nabla H_0$ in O^1.

Let us consider the equation

$$\dot{y} = JAy + V(y) . \tag{3.4}$$

Its solution $y(t)$ is the solution of equation (3.1) as long as $y(t) \in O^1$. Let us also consider the linear equation

$$\dot{y} = JAy . \tag{3.5}$$

As the operator JA is antisymmetric in Y_s, then by repeating the proof of Lemma 1.1, one obtains that the bounded operator $(JA)^{-1}$ is antiselfadjoint in Y_s. Thus the operator JA in Y_s with the domain of definition, equal to Y_{s+d_1} (i.e., equal to the image of $(JA)^{-1}$), is also antiselfadjoint. Due to the Stone's theorem [RS], for $y(0) = y_0 \in Y_{s+d_1}$ equation (3.5) has the unique strong solution and the flow of this equation defines isometries of the space Y_s. Equation (3.4) is a Lipschitz perturbation of (3.5). So it has the unique strong solution $y(t)$, $t \geq 0$, for every $y(0) \in Y_{s+d_1}$ and the unique weak solution for every $y(0) \in Y_s$ (see e.g., [Bre]). If $y(0) = y_0 \in O^2$, then due to the theorem's hypotheses, such a solution does not leave domain O^1 for $0 \le t \le T$ and for such a "t" it is the unique solution of equation (3.4). □

The theorem above reduces the problem of solving equation (3.1) to the problem of finding *a priori* estimate for its solutions.

4. Toroidal phase space

Let us consider a toroidal phase space $\mathcal{Y}$ of the form $\mathcal{Y} = \mathrm{T}^n \times \mathrm{R}^n \times Y$. Here $\mathrm{T}^n = \mathrm{R}^n / 2\pi \mathrm{Z}^n$ is the n-torus, $Y = Y_0$ and $\{Y_s | s \in \mathrm{R}\}$ is a Hilbert scale. Let us denote $\mathcal{Y}_s = \mathrm{T}^n \times \mathrm{R}^n \times Y_s$. Every space $\mathcal{Y}_s$ has the natural metric dist_s and the natural structure of a Hilbert manifold with the local charts

$$K(q^0) \times \mathrm{R}^n \times Y_s, \quad K(q^0) = \{q \in \mathrm{R}^n \mid |q_j - q_j^0| < \pi \ \forall j\}, \ q^0 \in \mathrm{T}^n$$

(see [Ch-B]). So

$$T_u \mathcal{Y}_s \cong \mathrm{R}^n \times \mathrm{R}^n \times Y_s =: Z_s \ \forall u \in \mathcal{Y}_s .$$

Let J^Y be an isomorphism of the scale $\{Y_s\}$ with properties a), b) and J^T be the map of $\mathrm{R}^n \times \mathrm{R}^n$ into itself, which maps a point (q, p) into the point $(p, -q)$. Let us denote by $J^{\mathcal{Y}}$ the operator

$$J^{\mathcal{Y}} = J^T \times J^Y : Z_s = (\mathrm{R}^n \times \mathrm{R}^n) \times Y_s \longrightarrow Z_{s-d_J} = (\mathrm{R}^n \times \mathrm{R}^n) \times Y_{s-d_J}$$

and introduce in $\mathcal{Y}_s$, $s \ge 0$, the 2-form

$$\alpha^{\mathcal{Y}} = \langle \bar{J}^{\mathcal{Y}} du, du \rangle_Z , \ \bar{J}^{\mathcal{Y}} = -(J^{\mathcal{Y}})^{-1} .$$

Definition. The triple $\{\mathcal{Y}, \{\mathcal{Y}_s\}, \alpha^{\mathcal{Y}}\}$ is called a toroidal symplectic Hilbert scale (TSHS).

Let O be a domain in $\mathcal{Y}_s$ and H be a C^1-function on O. The Hamiltonian equations corresponding to H have the form

$$\dot{q}_j = \frac{\partial H}{\partial p_j} , \ \dot{p}_j = -\frac{\partial H}{\partial q_j} \ (1 \le j \le n) , \ \dot{y} = J^Y \nabla_y H . \qquad (4.1)$$

The definitions of strong and weak solutions for equation (4.1) are analogous to those for equation (1.3).

The Poisson bracket of two C^1-functions H_1, H_2 on the domain O, such that $\nabla_y H_j \in C(O; Y_{s_j})$ $(j = 1, 2)$, where $s_1 + s_2 \ge d_J$, takes the form

$$\{H_1, H_2\}(q, p, y) = \sum_{j=1}^{n} \Big[-\frac{\partial H_1}{\partial q_j} \frac{\partial H_2}{\partial p_j} + \frac{\partial H_1}{\partial p_j} \frac{\partial H_2}{\partial q_j} \Big] + \langle J^Y \nabla_y H_1, \nabla_y H_2 \rangle_Y .$$

The results of section 1-3 readily extend to canonical transformations and Hamiltonian equations in TSHS. We formulate the analogs of Theorems 2.2, 2.4, 2.5 and 3.1 only.

Proposition 4.1. The statements of Theorem 2.2 remain true if anyone of the spaces X, Y is replaced by a toroidal symplectic Hilbert space (with equations of motion replaced accordingly).

Let $O^1 \subset O$ be domains in $\mathcal{Y}_s$ such that

$$\operatorname{dist}_s(O^1; \mathcal{Y}_s \backslash O) \geq \delta > 0 . \tag{4.2}$$

Let $H \in C^2(O)$ and

$$V_H = (\nabla_p H, -\nabla_q H, J^Y \nabla_y H)$$

be the corresponding Hamiltonian vector-field. Let us suppose that $V_H \in C^1(O; Z_s)$ and

$$|V_H(q,p,y)| \leq K \ \ \forall (q,p,y), \quad \operatorname{Lip}(V_H : O \longrightarrow Z_s) \leq K . \tag{4.3}$$

Then the flow mappings $S^t : O^1 \longrightarrow O$ exist for $0 \leq t \leq T = \delta/K$ and every map S^t is a C^1-diffeomorphism on its image.

Proposition 4.2. For every $0 \leq t \leq \delta/K$ the mapping S^t as above is a canonical transformation.

Let conditions (4.2), (4.3) be fulfilled and the maps $S^t \in C^1(O^1; O)$ form the flow of the equation

$$\frac{d}{dt}(q,p,y) = \varepsilon V_H(q,p,y) .$$

Proposition 4.3 If $G \in C^1(O)$, then

$$G(S^{t+\tau}(\mathfrak{h})) = G(\mathfrak{h}) + \tau\varepsilon\{H,G\}(S^t\mathfrak{h}) + O(\varepsilon\tau)^2$$

for $\mathfrak{h} = (q,p,y) \in O^1$ and $0 \leq t \leq t+\tau \leq T = \delta/K$. In particular, $\frac{d}{dt}G(S^t\mathfrak{h}) = \varepsilon\{H,G\}(S^t(\mathfrak{h}))$.

Let in (4.1) $H = \frac{1}{2}\langle Ay, y\rangle + H_0(q,p,y)$ and let the linear operator A be the same as in part 3. Let $O^1 \subset O^2 \subset O$ be domains in $\mathcal{Y}_s$, satisfying the inequality (4.2). Let us suppose that $\operatorname{Lip}(V_{H_0} : O \longrightarrow Z_s) \leq K$.

Proposition 4.4. If the operator JA is antisymmetric in Y_s and every strong solution of (4.1) with an initial point $\mathfrak{h}_0 = (q_0, p_0, y_0) \in O^2$ stays in domain O^1 for $0 \leq t \leq T$, then for $\mathfrak{h}_0 \in O^2 \cap \mathcal{Y}_{s+d_1}$ ($d_1 = d_A + d_J$) and for $0 \leq t \leq T$ there exists the unique strong solution of (4.1); for $\mathfrak{h}_0 \in O^2$, $0 \leq t \leq T$, there exists the unique weak solution of (4.1).

The proofs of Propositions 4.1–4.3 are the same as the proofs of the corresponding theorems.

5. A version of the former constructions

All the constructions of the sections 1–4 have the natural analogs for the scales of Hilbert spaces depending on an integer index, i.e. for the scales $\{Z_s | s \in \mathbb{Z}\}$. SHS and TSHS with discrete scales $\{Z_s\}$ are sometimes more convenient to study Hamiltonian equations of form (3.1) with integer d_A, d_J. For example, KdV equation (1.5), (1.5′) ($d_J = 1, d_A = 2$) and nonlinear Schrödinger equation (1.4) ($d_J = 0, d_A = 2$).

All the statements of sections 1–4 have natural analogs for discrete scales. The proofs are the same.

Part 2

Statement of the Main Theorem and its Consequences

The following notations are used: for domains O_1, O_2 in Banach spaces B_1, B_2 and for a Lipschitz map $h : O_1 \to O_2$ we denote

$$|h|_{B_2}^{O_1,\mathrm{Lip}} = \max\{ \sup_{b\in O_1} |h(b)|_{B_2}, \mathrm{Lip}\, h\} . \tag{0.1}$$

By B_1^c, B_2^c we denote the complexifications of the spaces B_1, B_2, i.e., $B_1^c = B_1 \otimes \mathbb{C}$ and the same for B_2^c. We suppose that the complex spaces are given norms, consistent with the norms in the real subspaces B_1, B_2, and denote these norms in the same way as the norms in B_1, B_2. (Such norms trivially exist if B_1, B_2 are Hilbert spaces or some L_p-spaces).

Let $O_1^c \subset B_1^c$, $O_2^c \subset B_2^c$ be open domains. We denote by $\mathcal{A}^R(O_1^c; O_2^c)$ the set of Fréchet complex-analytic maps from O_1^c to O_2^c, which are real for real arguments (i.e. map $O_1^c \cap B_1$ into B_2). Let M be some metric space. We denote by $\mathcal{A}_M^R(O_1^c; O_2^c)$ the class of the mappings $G : O_1^c \times M \to O_2^c$ which are analytic in the first variable and Lipschitz in the second variable. More explicit, $\mathcal{A}_M^R(O_1^c; O_2^c)$ consists of the maps G such that

i) $G(\cdot; m) \in \mathcal{A}^R(O_1^c; O_2^c)\ \forall m \in M$;
ii) for each b in O_1^c the map $G(b; \cdot) : M \to O_2^c$ is Lipschitz and

$$|G|_{B_2}^{O_1^c;M} \equiv \sup_{b\in O_1^c} |G(b;\cdot)|_{B_2}^{M,\mathrm{Lip}} < \infty . \tag{0.2}$$

For abstract sets $\mathfrak{A}$, $\mathcal{J}$, for a subset Θ of their product $\mathfrak{A} \times \mathcal{J}$ and for $I \in \mathcal{J}$ we denote by $\Theta[I]$ a subset of $\mathfrak{A}$ of the form

$$\Theta[I] = \{a \in \mathfrak{A} | (a, I) \in \Theta\} . \tag{0.3}$$

For a subset Q of a metric space M we denote by $O(Q, \delta, M)$ the δ-neighborhood of Q in M:

$$O(Q, \delta, M) = \{m \in M | \mathrm{dist}(m, Q) < \delta\} , \tag{0.4}$$

where $\mathrm{dist}(m, Q) = \inf_{m' \in Q} \mathrm{dist}(m, m')$; for a Banach space B we denote by $O(\delta, B)$ the δ-ball in B centered at zero.

For a Hilbert scale $\{Z_s | s \in \mathbb{R}\}$ we denote by $\|\cdot\|_{s_1,s_2}$ the operator norm in the space $\mathcal{L}(Z_{s_1}, Z_{s_2})$ of bounded linear operators $Z_{s_1} \to Z_{s_2}$.

In the notations of functions and mappings we sometimes omit part of the arguments; we denote by C, C_1, C_2 etc. different positive constants which arise in estimates and denote by K, K_1 etc. constants in the assumptions of theorems.

1. Statement of the main theorem

Let $\{Z,\{Z_s|s\in R\},\alpha=\langle \bar{J}^Z dz,dz\rangle_Z\}$ be a symplectic Hilbert scale as it was defined in Part 1. So $\bar{J}^Z=-(J^Z)^{-1}$ is an antiselfadjoint in Z bounded operator and the operator J^Z defines an isomorphism of the scale $\{Z_s\}$ of order $d_J\ge 0$. Let us suppose that the operator $\bar{J}^Z$ depends on a vector-parameter $a\in\mathfrak{A}$, where $\mathfrak{A}$ is a bounded connected open n-dimensional domain. So the symplectic form α also depends on the parameter a. Let $A^Z(a)$ be a selfadjoint operator in Z (possibly, unbounded), depending on $a\in\mathfrak{A}$, such that $A^Z(a)$ defines an isomorphism of the scale $\{Z_s\}$ of order $d_A\ge 0$. We suppose that for all a and s

$$\|J^Z(a)\|_{s,s-d_J},\|\bar{J}^Z(a)\|_{s,s+d_J},\|A^Z(a)\|_{s,s-d_A},\|(A^Z(a))^{-1}\|_{s,s+d_A}\le K\,, \quad (1.0)$$

and that there exists a basis $\{\varphi_j^\pm|j\ge 1\}$ of the space Z with the following properties:

i) there exist positive numbers $\lambda_j^{(s)}$, $s\in R$, $j\in N$, such that $\lambda_j^{(-s)}=(\lambda_j^{(s)})^{-1}$ for all j,s and

$$K^{-1}j^s\le\lambda_j^{(s)}\le Kj^s\,, \quad (1.1)$$

and $\{\varphi_j^\pm\lambda_j^{(-s)}|j\ge 1\}$ is a Hilbert basis of the space Z_s, i.e.,

$$\langle\varphi_j^{\sigma_1}\lambda_j^{(-s)},\varphi_k^{\sigma_2}\lambda_k^{(-s)}\rangle_s=\delta_{j,k}\,\delta_{\sigma_1,\sigma_2}\quad\forall\, j,k\in N\,,\ \forall\,\sigma_1,\sigma_2=\pm\,;$$

ii) for all j and a

$$J^Z(a)\varphi_j^\pm=\mp\lambda_j^J(a)\varphi_j^\mp\,, \quad (1.2)$$

$$A^Z(a)\varphi_j^\pm=\lambda_j^A(a)\varphi_j^\pm\,, \quad (1.3)$$

where the real numbers λ_j^J,λ_j^A are positive for j large enough.[8)]

Example. In Example 1.4 from Part 1 (see pp. 3, 5) let us take $A=-\partial^2/\partial x^2$ and supply $Z_0=L_2(\mathrm{T}^1)$ with the usual trigonometric basis

$$\varphi_j^+=\pi^{-1/2}\cos jx\,,\quad \varphi_j^-=\pi^{-1/2}\sin jx\quad (j=1,2,\dots)\,.$$

Then the functions $\{(1+j^2)^{-s/2}\varphi_j^\pm\}$ form a Hilbert basis of the space Z_s (so i) holds with $\lambda_j^{(s)}=(1+j^2)^{s/2}$) and the relations (1.2), (1.3) hold with $\lambda_j^J=j$, $\lambda_j^A=j^2$.

For examples of parameter-depending Hilbert scales as above see §§3, 5, 6 below.

Let us consider a hamiltonian

$$\mathcal{H}(z;a,\varepsilon)=\frac{1}{2}\langle A^Z(a)z,z\rangle+\varepsilon H(z;a,\varepsilon)\,, \quad (1.4)$$

depending on the parameter $a\in\mathfrak{A}$ and a small parameter $\varepsilon\in[0,1]$. The corresponding Hamiltonian equation (with respect to the symplectic structure given by the 2-form $\alpha(a)$) has the form

$$\dot{z}=J^Z(a)\big(A^Z(a)z+\varepsilon\nabla H(z;a,\varepsilon)\big)\,. \quad (1.5)$$

---

8) The assumption ii) may be essentially weakened. See §7 below.

Here and in what follows, ∇ is the gradient in $z \in Z$ with respect to the scalar product $\langle \cdot, \cdot \rangle$. Equation (1.5) is a perturbation of the linear Hamiltonian equation

$$\dot{z} = J^Z(a)A^Z(a)z \,. \tag{1.6}$$

The operator $J^Z A^Z$ defines an isomorphism of the scale of order $d_1 = d_A + d_J$. In view of conditions (1.2), (1.3) it is antiselfadjoint in Z (with the domain of definition Z_{d_1}) and its spectrum is pure imaginary,

$$\sigma\big(J^Z(a)A^Z(a)\big) = \{\pm i\lambda_j(a) | j \ge 1\}\,, \quad \lambda_j(a) = \lambda_j^J(a)\lambda_j^A(a)\,.$$ [9]

It is supposed that the functions $\lambda_j^J(a)$, $\lambda_j^A(a)$, $1 \le j \le n$, are C^2-smooth,

$$|\partial_a^\alpha \lambda_j^J(a)| + |\partial_a^\alpha \lambda_j^A(a)| \le K_1\,, \quad |\alpha| \le 2\,, \tag{1.7}$$

and the mapping $a \mapsto \omega = (\lambda_1, \ldots, \lambda_n) \in \mathbb{R}^n$ is nondegenerate at some point $a_0 \in \mathfrak{A}$:

$$\big|\det\{(\partial\omega_j/\partial a_k)(a_0) | 1 \le j,k \le n\}\big| \ge K_0 > 0\,. \tag{1.8}$$

We use special notations for the values of λ_j^A, λ_j^J, λ_j and ω at the point a_0 and denote

$$\lambda_{j0}^A = \lambda_j^A(a_0)\,,\ \lambda_{j0}^J = \lambda_j^J(a_0)\,,\ \lambda_{j0} = \lambda_j(a_0)\,,\ \omega_0 = \omega(a_0)\,.$$

Let us set $Z^0 \subset Z$ be equal to the $2n$-dimensional linear span of the vectors $\{\varphi_j^\pm \mid j \le n\}$. The space Z^0 is foliated into tori $T(I)$ of dimension $\le n$, which are invariant for linear equation (1.6),

$$T(I) = \{\sum_{j=1}^n \alpha_j^+\varphi_j^+ + \alpha_j^-\varphi_j^- \mid {\alpha_j^+}^2 + {\alpha_j^-}^2 = 2I_j \ge 0 \ \forall j\}\,.$$

A torus $T(I)$ with all I_j positive is n-dimensional. Let

$$\Sigma_I^0 : \mathrm{T}^n \longrightarrow T^n(I) \subset Z^0 \tag{1.9}$$

be the natural parametrisation of $T^n(I)$ i.e., Σ_I^0 is the map which identify a point $(q_1, \ldots, q_n)$ from the standard n-torus $\mathrm{T}^n = \mathbb{R}^n/2\pi\mathbb{Z}^n$ with the point in $T^n(I)$ such that $\alpha_j^+ = \cos q_j$, $\alpha_j^- = \sin q_j$ $\forall j$. Each torus $T^n(I)$ is filled with quasiperiodic solutions of the linear equation (1.6) of the form $t \longmapsto \Sigma_I^0(q_0 + \omega t)$.

Let us consider a family of the tori $\{T(I) | I \in \mathcal{J}\}$, where $\mathcal{J}$ is a Borel set which lies in a compact part of the octant $\mathbb{R}_+^n$:

$$\mathcal{J} \subset \{I \in \mathbb{R}^n | K_1^{-1} \le I_j \le K_1 \ \forall j\}\,.$$

Possibly, $\mathcal{J}$ consists of the only point I_0 (i.e., $\mathcal{J} = \{I_0\}$). Let us denote by $\mathcal{T}$ the union of all the tori $T(I)$,

$$\mathcal{T} = \cup\{T(I) | I \in \mathcal{J}\}\,,$$

[9] In fact, a finite subset of the eigenvalues of the operator $J^z(a)A^z(a)$ could have nontrivial real parts. See Part 2.7 and Refinement 3 in Part 3.1.

fix some number $d \geq d_A/2$ and denote by O^c complex K_1^{-1}-neighborhood of T in Z_d^c, $O^c = O(K_1^{-1}, T, Z_d^c)$ (see (0.4) for this notation). We suppose that for all ε the function $H(z; a, \varepsilon)$ may be extended to a function $H : O^c \times \mathfrak{A} \times [0,1] \to \mathbb{C}$ which is complex-analytic in $z \in O^c$ and Lipschitz in $a \in \mathfrak{A}$.

In what follows we use the map

$$\Sigma^0 : \mathrm{T}^n \times \mathfrak{A} \times \mathcal{J} \longrightarrow Z\ , \ (q, a, I) \mapsto \Sigma_I^0(q)\ .$$

Theorem 1.1. Let the conditions mentioned above hold together with

1) *analyticity and quasilinearity*: for some $d_H \in \mathbb{R}$ such that

$$d_H < d_A - 1\ ,\ d_J + d_H \leq 0\ , \tag{1.10}$$

and for all $\varepsilon \in [0,1]$

$$|H(\cdot;\cdot,\varepsilon)|^{O^c;\mathfrak{A}} \leq K_1\ ,\quad \|\nabla H(\cdot;\cdot,\varepsilon)\|_{d_c}^{O^c;\mathfrak{A}} \leq K_1\ ,$$

where $d_c = d - d_H$;

2) *spectral asymptotics*: $d_1 \equiv d_A + d_J \geq 1$ and there exists an asymptotic expansion for the frequencies λ_{j0} as $j \to \infty$:

$$\left|\lambda_{j0} - K_2^0 j^{d_1} - K_2^1 j^{d_{1,1}} - \ldots - K_2^{r-1} j^{d_{1,r-1}}\right| \leq K_1 j^{d_{1,r}} \quad r \geq 1\ , \tag{1.11}$$

with positive $K_2^0 > 0$, real K_2^j and with the numbers $d_{1,j}$ such that $d_1 > d_{1,1} > \cdots > d_{1,r}$, $d_1 - 1 > d_{1,r}$;

$$\mathrm{Lip}\lambda_j^J \leq K_1 j^{d_J}\ ,\ \mathrm{Lip}\lambda_j \leq K_1 j^{d_{1,r}}\ . \tag{1.12}$$

Then there exist large enough integers j_1, M_1 such that if the condition

3) *nonresonance*:

$$\begin{gathered} |\ell_1\lambda_{10} + \ell_2\lambda_{20} + \cdots + \ell_{j_1}\lambda_{j_1 0}| \geq K_3 > 0 \quad \text{for all} \quad \ell \in \mathbb{Z}^{j_1} \\ \text{such that} \quad |\ell| \leq M_1 \quad \text{and} \quad 1 \leq |\ell_{n+1}| + \cdots + |\ell_{j_1}| \leq 2 \end{gathered} \tag{1.13}$$

is satisfied, then there exists $\delta_a > 0$ and for arbitrary $\gamma > 0$ and for $\varepsilon \in \big(0, \varepsilon(\gamma)\big)$ with small enough positive $\varepsilon(\gamma)$ there exists an ε-dependent Borel set Θ_ε of vectors (a, I),

$$\Theta_\varepsilon \subset \Theta^{a_0} = O(a_0, \delta_a, \mathfrak{A}) \times \mathcal{J}\ ,$$

and analytic embeddings

$$\Sigma_{(a,I)}^\varepsilon : \mathrm{T}^n \longrightarrow Z_{d_c}\ ,\quad (a, I) \in \Theta_\varepsilon\ , \tag{1.14}$$

with the following properties a) – d):

a) $\mathrm{mes}(\Theta^{a_0} \backslash \Theta_\varepsilon)[I] \leq \gamma$ for all $I \in \mathcal{J}$ (see (0.3));

b) the mapping

$$\Sigma^\varepsilon : \mathrm{T}^n \times \Theta_\varepsilon \longrightarrow Z_{d_c}\,, \quad (q,a,I) \longmapsto \Sigma^\varepsilon_{(a,I)}(q)$$

is Lipschitz-close to the mapping Σ^0:

$$\|\Sigma^\varepsilon - \Sigma^0\|_{d_c}^{T^n\times\Theta,\mathrm{Lip}} \le C(\gamma)\varepsilon\,;$$

c) every torus $\Sigma^\varepsilon_{(a,I)}(\mathrm{T}^n)$, $(a,I)\in\Theta_\varepsilon$, is invariant for the equation (1.5) and is filled with weak in Z_d quasiperiodic solutions of the form $z^\varepsilon(t) = \Sigma^\varepsilon_{(a,I)}(q+\omega' t)$, where $q\in\mathrm{T}^n$ and $\omega' = \omega'(a,I,\varepsilon)$ is an n-vector such that $|\omega-\omega'|\le C(\gamma)\varepsilon$;

d) all Lyapunov exponents of the solutions $z^\varepsilon(t)$ are equal to zero.

The theorem will be proved and somewhat refined in Part 3.2. See there Theorem 3.11, Remarks 2, 3, 4 and Refinement 3.

Let ρ be any positive number, smaller that one. We can apply the theorem with $\gamma = 1/2, 1/3, \ldots$ and obtain the sets $\Theta_{n,\varepsilon}\subset\Theta^{a_0}$ and positive numbers ε_n such that

$$\mathrm{mes}(\Theta^{a_0}\backslash\Theta_{n,\varepsilon})[I] \le 1/n\,,$$

and

$$\|\Sigma^\varepsilon - \Sigma^0\|_{d_c}^{T^n\times\Theta_{n,\varepsilon},\mathrm{Lip}} \le \varepsilon^\rho,\ |\omega-\omega'|\le\varepsilon^\rho\,, \tag{1.15}$$

provided that $\varepsilon\le\varepsilon_n$. We may suppose that the sequence $\{\varepsilon_n\}$ monotonically tends to zero, choose $n = n(\varepsilon) = \sup\{n|\varepsilon_n > \varepsilon\}$ and denote $\Theta_\varepsilon = \Theta_{n(\varepsilon),\varepsilon}$. We have obtained the following consequence of the theorem:

Corollary 1.1. Under the assumptions 1) – 3) of Theorem 1.1, there exists $\delta_a > 0$ and for arbitrary $\rho < 1$ and ε small enough, there exists a Borel set $\Theta_\varepsilon\subset\Theta^{a_0}$, for $(a,I)\in\Theta_\varepsilon$ exists a vector $\omega'(a,I,\varepsilon)$ and analytic embedding $\Sigma^\varepsilon_{(a,I)}$ as in (1.14), such that

a) $\mathrm{mes}(\Theta^{a_0}\backslash\Theta_\varepsilon)[I]\longrightarrow 0$ when $\varepsilon\to 0$, uniformly with respect to I;

b) the mapping Σ^ε is Lipschitz and the estimate (1.15) holds with $\Theta_{n,\varepsilon} = \Theta_\varepsilon$;

c) the tori $\Sigma^\varepsilon_{(a,I)}(\mathrm{T}^n)$, $(a,I)\in\Theta_\varepsilon$, are invariant for the equation (1.5) and are filled with weak in Z_d solutions of the form $t\longmapsto\Sigma^\varepsilon_{(a,I)}(q+\omega' t)$. All Lyapunov exponents of these solutions are equal to zero.

Another immediate consequence of Theorem 1.1 is a strong averaging principle for nonresonant systems of the form (1.5):

Corollary 1.2. Under the assumptions of Theorem 1.1 for every $(a,I)\in\Theta_\varepsilon$, $q\in\mathrm{T}^n$, and for all t the curve $t\longmapsto\Sigma^0_{(a,I)}(q+\omega' t)$ for ε small enough is $C\varepsilon$-close to some weak solution of (1.5). Here ω' is "an averaged frequency vector", $|\omega'-\omega|\le C\varepsilon$.

Remarks. 1) By the estimate on ∇H in the assumptions of the theorem, the order of nonlinear operator in equation (1.5) does not exceed $d_J + d_H$. The order of the linear one is equal to $d_J + d_A$. So the assumption (1.10) of Theorem 1.1 indeed means the quasilinearity of equation (1.5) because the order of the linear term exceeds the order of the nonlinear term more than by one.

2) If $d_0 \le d_c - d_1$, then the r.h.s. in (1.5) with $z(t) = z^\varepsilon(t)$ defines a continuous curve in Z_{d_0}. So $z^\varepsilon \in C^1([0,T]; Z_{d_0})$ is a strong in Z_{d_0} solution of (1.5).

3) The numbers j_1, M_1, δ_a in the assumption 3) of Theorem 1.1 depend on K, $K_0, K_1, K_2^j, d_{1,j}, d_A, d_J, d_H$, d, n only. The function $\gamma \to \varepsilon(\gamma)$ depends on the same quantities and on K_3 from (1.13).

4) All the tori $\Sigma^\varepsilon_{(a,I)}(\mathrm{T}^n)$ are isotropic, i.e. $\left[\Sigma^\varepsilon_{(a,I)}\right]^* \alpha = 0$. The reason is that in the proof of the theorem we give below the map $\Sigma^\varepsilon_{(a,I)}$ is constructed as the limit (in the C^∞-topology) of isotropic imbeddings of the n-torus into Z.

5) The frequencies $\{\lambda_{j0}\}$ are ordered asymptotically only. So for a space Z^0 one can choose any $2n$-dimensional invariant subspace of the operator $J(a)A(a)$.

6) If instead of the condition $d_1 \ge 1$ a weaker condition $d_1 > 0$ holds, then the statements of Theorem 1.1 seems to be wrong in a general case. But the statement of Corollary 1.2 remains true for $0 \le t \le \varepsilon^{-1}$ with $C\varepsilon^\rho$ instead of $C\varepsilon$ and some positive ρ (see [K4]).

7) The form of the spectral condition in the theorem's assumptions is not the most general one we need for the proof. For example (see [K1]), for $d_1 > 1$ it is sufficient to suppose that

$$C^{-1}j^{d_1} - C \le \lambda_j \le Cj^{d_1} + C,\ |\lambda_{j+1} - \lambda_j| \ge C_1 j^{d_1-1}\ \forall j\,.$$

See also [K2] for a possible form of the spectral condition with $d_1 = 1$. For an investigation of the problem see [DPRV].

8) The analyticity of the invariant tori $\Sigma^\varepsilon_{(a,I)}(\mathrm{T}^n)$ was observed by J. Pöschel [P1]. In the author's works [K1–K3] only smoothness of the tori was stated.

9) If all the numbers d, d_H, d_A, d_J are integers, then Theorem 1.1 may be stated in the framework of discrete symplectic Hilbert scales $\{Z, \{Z_s | s \in \mathbb{Z}\}, \alpha\}$ (see Part 1.5).

2. Reformulation of Theorem 1.1

Let $\mathfrak{A}$ be a closed connected n-dimensional domain with the piecewise smooth boundary $\partial\mathfrak{A}$; let us suppose that λ_j^J, λ_j^A are analytic functions of the parameter $a \in \mathfrak{A}$ (i.e., may be analytically extended to a neighbourhood of $\mathfrak{A}$), and the vector $\omega(a) = (\lambda_1, \dots, \lambda_n)(a)$ nondegenerately depends on a:

$$\det\{\partial\omega_j/\partial a_k | 1 \le j, k \le n\} \not\equiv 0\,. \tag{2.1}$$

Let us consider an identical resonance relation of the form

$$s \cdot \omega(a) + \Lambda(a) \equiv 0\,,\ \Lambda = \ell_1\lambda_{n+1}(a) + \dots + \ell_p\lambda_{n+p}(a)\,, \tag{2.2}$$

with some integer n-vector s and p-vector ℓ such that

$$1 \le |\ell_1| + |\ell_2| + \cdots + |\ell_p| \le 2 . \tag{2.3}$$

Lemma 2.1. Let all the functions λ_j^J, λ_j^A be analytic in $\mathfrak{A}$ and asymptotics (1.11) with $d_1 \ge 1$ hold together with assumptions (1.12), (2.1). Then there exist numbers M_2, j_2 with the following property: if some identical relation of the form (2.2), (2.3) holds with $\ell_p \ne 0$, then $|s| \le M_2$ and $p \le j_2$.

Proof. By the assumption (2.1), there exist a point $a' \in \mathfrak{A}$ and a constant C such that

$$C^{-1} \le |\omega_*(a')|_{\mathbf{R}^n,\mathbf{R}^n} \le C . \tag{2.4}$$

Let ∇_a be the gradient in a with respect to the usual scalar product in $\mathbf{R}^n$. Then $\nabla_a(s \cdot \omega(a)) = \omega^*(a)s$ and by (2.4)

$$\left|\nabla_a\big(s \cdot \omega(a')\big)\right| \ge C^{-1}|s| . \tag{2.5}$$

Till the end of the proof we denote by $\lambda_n(p)$ zero function: $\lambda_n(p) \equiv 0$. Then every relation of the form (2.2), (2.3) with $\ell_p \ne 0$ may be rewritten in the following way:

$$s' \cdot \omega(a) \pm \lambda_m(a) + \lambda_{n+p}(a) \equiv 0 . \tag{2.6}$$

Here $s' = s$ or $s' = s/2$ (if in (2.2) $|\ell_j| = 2$ for some $n+1 \le j \le n+m$) and $n \le m < n+p$. It follows from (2.6) that $|\lambda_{n+p}(a) \pm \lambda_m(a)| \le C_1|s|$. Thus by the assumption (1.11)

$$C_2|s| \ge (n+p)^{d_1} - (n+p)^{d_{1,r}} - m^{d_1} - m^{d_{1,r}} . \tag{2.7}$$

It follows from (2.5), (2.6) that $C_3^{-1}|s| \le |\nabla\lambda_m(a) \pm \nabla\lambda_{n+p}(a)|$. So the assumption (1.12) implies the estimate for $|s|$ in terms of m and $n+p$:

$$C_3\big((n+p)^{d_{1,r}} + m^{d_{1,r}}\big) \ge |s| . \tag{2.8}$$

The last two estimates jointly imply that

$$C\big((n+p)^{d_{1,r}} + m^{d_{1,r}}\big) \ge (n+p)^{d_1} - m^{d_1} .$$

Since $n+p \ge m+1$ and the function t^{d_1} is convex, it follows that

$$\frac{1}{2}\big((n+p)^{d_1} - m^{d_1}\big) \ge C_1^{-1}(n+p)^{d_1-1}$$

and

$$\frac{1}{2}\big((n+p)^{d_1} - m^{d_1}\big) \ge C_1^{-1}(n+p-m)m^{d_1-1} .$$

By the last three estimates, $C_2\big((n+p)^{d_{1,r}} + m^{d_{1,r}}\big) \ge (n+p)^{d_1-1} + (n+p-m)m^{d_1-1}$. So $n+p \le C'$ (to prove the estimate one has to use the inequality $d_{1,r} < d_1 - 1$ and to consider the cases $d_1 > 1$ and $d_1 = 1$ separately). Now by (2.8) $|s| \le C''$ and the lemma is proved with $j_2 = C' - n$, $M_2 = C''$. □

Theorem 2.2. Let all the eigen-values λ_j^A, λ_j^J be analytic functions of the parameter $a \in \mathfrak{A}$ and the assumption (2.1) holds together with assumptions 1), 2) of Theorem 1.1. Then there exist integers j_1, M_1 such that if

$$s_1\lambda_1(a) + s_2\lambda_2(a) + \cdots + s_{j_1}\lambda_{j_1}(a) \not\equiv 0 \tag{2.9}$$

for all integer j_1-vectors s such that

$$|s| \le M_1\,,\ 1 \le |s_{n+1}| + \cdots + |s_{j_1}| \le 2\,, \tag{2.10}$$

then for every $\gamma > 0$ and for sufficiently small $\varepsilon > 0$ there exists a Borel subset $\Theta_\varepsilon \subset \Theta_0 \equiv \mathfrak{A} \times \mathcal{J}$ and analytic embeddings

$$\Sigma^\varepsilon_{(a,I)} : \mathrm{T}^n \longrightarrow Z_{d_c}\,,\ (a, I) \in \Theta_\varepsilon\,,\ d_c = d - d_H\,,$$

with the following properties a) – c):

a) $\operatorname{mes}(\Theta_0 \backslash \Theta_\varepsilon)[I] < \gamma \ \ \forall I \in \mathcal{J}$,

b) the mapping

$$\Sigma^\varepsilon : \mathrm{T}^n \times \Theta_\varepsilon \longrightarrow Z_{d_c}\,,\ (q, a, I) \longmapsto \Sigma^\varepsilon_{(a,I)}(q)$$

is Lipschitz and

$$\left\|\Sigma^\varepsilon - \Sigma^0\right\|_{d_c}^{T^n \times \Theta_\varepsilon, \mathrm{Lip}} \le C_\gamma \varepsilon\,, \tag{2.11}$$

c) every torus $\Sigma^\varepsilon_{(a,I)}(\mathrm{T}^n)$, $(a, I) \in \Theta_\varepsilon$, is invariant for the equation (1.5) and is filled with weak in Z_d quasiperiodic solutions of the form $t \longmapsto \Sigma^\varepsilon_{(a,I)}(q + \omega' t)$, where $q \in \mathrm{T}^n$ and

$$|\omega - \omega'| \le C^1_\gamma \varepsilon\,, \tag{2.12}$$

d) all Lyapunov exponents of these solutions are equal to zero.

Proof. For short we consider the case of a one-point set $\mathcal{J} = \{I_0\}$ only. Now the set Θ_ε we are looking for has the form $\Theta_\varepsilon = \mathfrak{A}_\varepsilon \times \{I_0\}$.

By the analyticity of the functions λ_j^J, λ_j^A and by the assumption (2.1), the set $\{a \in \mathfrak{A} \mid |\det \partial\omega_j/\partial a_k| > 0\}$ is open and of full Lebesgue measure in $\mathfrak{A}$ (i.e., the measure of its complement is equal to zero). Let us define

$$\mathfrak{A}_t = \{a \in \mathfrak{A} \mid |\det \partial\omega_j/\partial a_k| \ge t\}\,.$$

Then the sets $\mathfrak{A}_t$ with $t \to 0$ form an increasing sequence of compact sets and the union $\cup\{\mathfrak{A}_t | t > 0\}$ is of full measure. So there exists $K_0 = K_0(\gamma) > 0$ such that

$$\operatorname{mes} \mathfrak{A} \backslash \mathfrak{A}_{K_0} < \gamma_0 = \gamma/4\,. \tag{2.13}$$

Let us choose in (2.9), (2.10) $j_1 \geq j_2 + n$, $M_1 \geq M_2 + 2$ with j_2, M_2 as in Lemma 2.1. Then by the assumption (2.9), (2.10) and Lemma 2.1 there is no identical resonance relation of the form (2.2), (2.3). So every set

$$\{a \in \mathfrak{A} | s_1\lambda_1(a) + \cdots + s_p\lambda_p(a) \neq 0\} \tag{2.14}$$

with $1 \leq |s_{n+1}| + \cdots + |s_p| \leq 2$, is of full measure in $\mathfrak{A}$.

We are going to apply Theorem 1.1 with $a_0 \in \mathfrak{A}_{K_0}$. By Remark 3, Theorem 1.1 is applicable with this choice of the point a_0, if condition (1.13) is fulfilled with some numbers j_1, M_1 which do not depend on a_0. Let us consider the set $\mathfrak{B}_t$ of all $a \in \mathfrak{A}$ such that

$$|s_1\lambda_1(a) + \cdots + s_{j_1}\lambda_{j_1}(a)| \geq t$$

for all integer j_1-vectors s, satisfying (2.10). As all the sets (2.14) are of full measure, then for some $t = K_3$

$$\operatorname{mes} \mathfrak{A} \setminus \mathfrak{B}_{K_3} < \gamma_0 \, . \tag{2.15}$$

Theorem 1.1 is applicable with arbitrary $a_0 \in \mathfrak{A}_{K_0} \cap \mathfrak{B}_{K_3}$, $\mathcal{J} = \{I_0\}$ and with the constant K_0 in the assumption (1.8) as in (2.13). In this situation, by Remark 3, the radius δ_a does not depend on a_0. The set Θ_ε has the form

$$\Theta_\varepsilon = \mathfrak{A}_\varepsilon^{a_0} \times \{I_0\} \, , \; \mathfrak{A}_\varepsilon^{a_0} \subset O(a_0, \delta_a, \mathfrak{A}) \tag{2.16}$$

and the open balls $O(a_0, \delta_a, \mathfrak{A})$ with $a_0 \in \mathfrak{A}_{K_0} \cap \mathfrak{B}_{K_3}$, form a covering of the compact set $\mathfrak{A}_{K_0} \cap \mathfrak{B}_{K_3}$.

Let us fix some finite subcovering $\{D_j = O(a_j, \delta_a, \mathfrak{A}) \mid j = 1, \ldots, M\}$ of this set. By the assertion a) of Theorem 1.1,

$$\operatorname{mes} D_j \setminus \mathfrak{A}_\varepsilon^{a_j} < \gamma_0 / M \quad \forall j = 1, \ldots, M \quad \text{if} \quad \varepsilon < \varepsilon(\gamma/4M) \, . \tag{2.17}$$

For every $j = 1, \ldots, M$ let us choose a closed subset $D_j^0 \subset D_j$ in such a way that

$$\operatorname{mes} (\cup D_j \setminus \cup D_j^0) < \gamma_0 \tag{2.18}$$

and

$$\operatorname{dist}(D_j^0, D_k^0) \geq \delta' > 0 \;\; \forall j \neq k \tag{2.19}$$

with some $\delta'(\gamma)$. Let us set

$$\mathfrak{A}_\varepsilon = \bigcup_{j=1}^{M} (\mathfrak{A}_\varepsilon^{a_j} \cap D_j^0)$$

and define the map Σ_a^ε with $a \in \mathfrak{A}_\varepsilon^{a_j} \cap D_j^0$, be equal to the map $\Sigma_{(a,I_0)}^\varepsilon$, constructed by means of Theorem 1.1 for $a \in \mathfrak{A}_\varepsilon^{a_j}$. This definition is correct because every point $a \in \mathfrak{A}_\varepsilon$ belongs to only one set $\mathfrak{A}_\varepsilon^{a_j} \cap D_j^0$.

The statements a) – c) of the theorem holds with this choice of $\mathfrak{A}_\varepsilon$ and Σ_a^ε. Indeed, the assertion a) results from the estimates (2.13), (2.15), (2.17), (2.18). The assertion c) is local with respect to the parameter a, so it results from Theorem

1.1. To prove the assertion b), let us observe that by Theorem 1.1 for the map $\Delta\Sigma = \Sigma^\epsilon - \Sigma^0$ we have the estimate

$$\|\Delta\Sigma\|_{d_c}^{T^n \times \mathfrak{A}_\epsilon^{a_j}, \mathrm{Lip}} \le C'_\gamma \epsilon \,. \tag{2.20}$$

If $b_j \in \mathfrak{A}_\epsilon^{a_j} \cap D_j^0$, $b_k \in \mathfrak{A}_\epsilon^{a_k} \cap D_k^0$ and $j \neq k$, then $|b_j - b_k| \ge \delta'$ by (2.19). This estimate and (2.20) jointly imply that

$$\|\Delta\Sigma(q; b_1) - \Delta\Sigma(q; b_2)\|_{d_c} \le 2C'\delta'^{-1} |b_1 - b_2| \epsilon \quad \forall b_1, b_2 \in \mathfrak{A}_\epsilon \,. \tag{2.21}$$

By (2.20), (2.21) we get the estimate (2.11) with $C_\gamma = C'(1 + 2\delta'^{-1})$. □

The following statement results from Theorem 2.2 in the same way as Corollary 1.1 results from Theorem 1.1.

Corollary 2.3. If under the assumptions of Theorem 2.2, assumption (2.9), (2.10) is satisfied, then for arbitrary $\rho \in (0,1)$ and for $0 < \epsilon < \epsilon(\rho)$ there exists a Borel subset $\Theta_\epsilon \subset \Theta_0$ and analytic embeddings $\Sigma^\epsilon_{(a,I)} : \mathrm{T}^n \to Z_{d_c}$, $(a, I) \in \Theta_\epsilon$, with the following properties:

a) $\mathrm{mes}(\Theta_0 \setminus \Theta_\epsilon)[I] \longrightarrow 0$ when $\epsilon \longrightarrow 0$, uniformly in I,

b) $\|\Sigma^\epsilon - \Sigma^0\|_{d_c}^{T^n \times \Theta_\epsilon, \mathrm{Lip}} \le \epsilon^\rho$,

c) every torus $\Sigma^\epsilon_{(a,I)}(\mathrm{T}^n)$, $(a, I) \in \Theta_\epsilon$, is invariant for the equation (1.5) and is filled with weak in Z_d quasiperiodic solutions of the form $\Sigma^\epsilon_{(a,I)}(q + \omega' t)$, $|\omega' - \omega| \le \epsilon^\rho$. All Lyapunov exponents of these solutions are equal to zero.

3. Nonlinear Schrödinger equation

The nonlinear Schrödinger equation

$$\dot u = i\Big(-u_{xx} + V(x; a)u + \epsilon \frac{\partial}{\partial |u|^2} \chi(x, |u|^2; a)u\Big) \,, \quad u = u(t, x) \,,$$

where the real valued functions V and χ depend on a parameter a from a bounded n-dimensional domain $\mathfrak{A}$, will be considered under the Dirichlet boundary conditions

$$0 \le x \le \pi \,, \quad u(t, 0) \equiv u(t, \pi) \equiv 0 \,.$$

Let $Z = L_2(0, \pi; \mathrm{C})$ which is regarded as a real Hilbert space with the inner product

$$\langle u, v\rangle = Re \int u(x)\overline{v(x)}\, dx \,.$$

The differential operator $-\partial^2/\partial x^2$ with the Dirichlet boundary conditions defines a positive selfadjoint operator $\mathcal{A}_0$ in Z with the domain of definition $D(\mathcal{A}_0) = (\overset{\circ}{H}{}^1 \cap H^2)(0, \pi; \mathrm{C})$. For $s \ge 0$ let Z_s be the domain of definition of the operator $\mathcal{A}_0^{s/2}$. Every space Z_s is a closed subspace of the Sobolev space $H^s(0, \pi; \mathrm{C})$ and the natural norm in Z_s is equivalent to the norm induced from $H^s(0,\pi;\mathrm{C})$. In particular,

$$Z_1 = \mathring{H}^1(0,\pi;\mathrm{C})\ , \quad Z_2 = (\mathring{H}^1 \cap H^2)(0,\pi;\mathrm{C})$$

(see e.g. [RS]). Let Z_{-s} be the Hilbert space adjoint to Z_s with respect to the scalar product in Z (see Part 1).

To obtain a symplectic Hilbert scale, let us take for the antiselfadjoint operator J the operator, mapping complex-valued function $u(x)$ to $iu(x)$. This operator gives an isomorphism of the scale $\{Z_s\}$ of order zero and $J^2 = -E$. So $\bar{J} = -(J^{-1}) = J$ and the triple $\{Z, \{Z_s\}, \langle J dz, dz\rangle\}$ is a symplectic Hilbert scale.

Let us suppose that the potential $V(x;a)$ is a bounded measurable function, Lipschitz in a:

$$\sup_a |V(x;a)| + \mathrm{Lip}(a \longmapsto V(x;a)) \leq K_1 \quad \forall x\ . \tag{3.1}$$

The differential operator $-\partial^2/\partial x^2 + V(x;a)$, $a \in \mathfrak{A}$, defines a selfadjoint operator $\mathcal{A}(a)$ in Z with the domain of definition Z_2. For a complete system of eigenvectors of $\mathcal{A}(a)$ let us take $\{\varphi_j^{\pm}(a)\}$, where $\varphi_j^{+}(a) = \varphi_j(x;a)$, $\varphi_j^{-}(a) = i\varphi_j(x;a)$ and $\{\varphi_j(x;a)\}$ is the complete in $L_2(0,\pi;\mathrm{R})$ system of real eigen-functions of the operator $-\partial^2/\partial x^2 + V(x;a)$ under the Dirichlet boundary conditions. So

$$\mathcal{A}(a)\varphi_j^{\pm}(a) = \lambda_j^{A}(a)\varphi_j^{\pm}(a)\ , \quad \varphi_j^{\pm} \in Z_2\ , \quad \|\varphi_j^{\pm}\|_0 = 1 \quad \forall j \geq 1\ .$$

The numbers $\{\lambda_j^{A}(a)\}$ are supposed to be asymptotically ordered, i.e. $\lambda_j^{A}(a) > \lambda_k^{A}(a)$ if $j > k$ and k is large enough.

Let $O^c \subset \mathrm{C}$ be a complex neighborhood of the real line R and let us suppose that the function χ may be extended into $[0,\pi] \times O^c \times \mathfrak{A}$ to a function which is C^2-smooth in x, analytic in u and Lipschitz in a:

$$\begin{aligned} &\chi(\cdot,\cdot;a) \in C^2([0,\pi] \times O^c;\mathrm{C}) \quad \forall a\ , \\ &\frac{\partial^s}{\partial x^s}\chi(x,\cdot;\cdot) \in \mathcal{A}_{\mathfrak{A}}^{R}(O_B^c;\mathrm{C}) \qquad \forall s \leq 2\ , \quad \forall x\ , \end{aligned} \tag{3.2}$$

for each bounded subset O_B^c of O^c.

Let us set

$$H_0(u;a) = \frac{1}{2}\int_0^{\pi} \chi(|u(x)|^2, x; a)\, dx\ . \tag{3.3}$$

Lemma 3.1. For any $R > 0$ there exists $\delta = \delta(R) > 0$ and a complex δ-neighborhood $B^c \subset Z_2^c$ of the ball $\{u \in Z_2 \mid \|u\|_2 \leq R\}$ such that the function H_0 is analytic there, $\nabla H_0 \in \mathcal{A}_{\mathfrak{A}}^{R}(B^c; Z_2^c)$ and

$$\nabla H_0(u;a) = \frac{\partial}{\partial |u|^2}\chi(x, |u|^2; a)u\ . \tag{3.4}$$

Proof. The existence of the set B^c and analyticity of H_0 result from Corollary A2 (see Appendix). The relation (3.4) results from the identities

$$\begin{aligned}\langle v(x), \nabla H_0(u(x);a)\rangle_Z &= dH_0(u;a)(v)\\ &= \frac{1}{2}\int_0^\pi \frac{\partial}{\partial |u|^2}\chi(x,|u|^2;a)(u\bar{v}+\bar{u}v)\,dx\\ &= \langle v(x), \frac{\partial}{\partial |u|^2}\chi(x,|u(x)|^2;a)u\rangle_Z .\end{aligned}$$

By Corollary A2, the map ∇H_0 belongs to $\mathcal{A}^R_{\mathfrak{A}}(B^c; H^2(0,\pi;\mathrm{C}))$. Besides, for $u(x) \in B^c$ the function $\nabla H_0(u(x);a)$ is equal to zero for $x = 0,\ \pi$. Therefore $\nabla H_0(u(x);a) \in (H^2 \cap \mathring{H}^1)(0,\pi;\mathrm{C}) = Z_2^c$ for $u \in B^c$ and the analyticity of ∇H_0 is also proved. □

Due to (3.4), the Hamiltonian equation with the hamiltonian $\frac{1}{2}\langle \mathcal{A}(a)u,u\rangle_Z + H_0(u;a)$ has the form

$$\dot{u} = i\left(-u_{xx} + V(x;a)u + \varepsilon\frac{\partial}{\partial |u|^2}\chi(x,|u|^2;a)u\right). \tag{3.5}$$

This equation is of the form (1.5), but the operators $\mathcal{A}(a)$ do not commute and the condition (1.3) is not satisfied. To apply the theorem we begin with the linear transformation U_a of the phase space, depending on the prameter a:

$$U_a : Z \longrightarrow Z\ , \quad z\varphi_j^{\pm}(x) \longmapsto z\varphi_j^{\pm}(x;a) \quad \forall z \in \mathrm{R} \quad \forall j\,.$$

Here $\varphi_j^+(x) = \sqrt{(2/\pi)}\sin jx$, $\varphi_j^-(x) = i\varphi_j^+(x)$. Observe that the operator U_a commutes with J.

Lemma 3.2. For every $a \in \mathfrak{A}$ the transformation U_a is canonical and orthogonal with respect to the scalar product $\langle\cdot,\cdot\rangle_Z$. For every a, a_1, $a_2 \in \mathfrak{A}$ and every $s \in [0,2]$ the following estimates hold:

$$|\lambda_j(a_1) - \lambda_j(a_2)| \le C\,|a_1 - a_2|\ , \tag{3.6}$$

$$\|U_{a_1} - U_{a_2}\|_{s,s} \le C_s\,|a_1 - a_2|\ , \tag{3.7}$$

$$\|U_a\|_{s,s} \le C_s'\,. \tag{3.8}$$

Here $\|\cdot\|_{s,s}$ is the operator norm for linear operators in Z_s.

Proof. The orthogonality of U_a results from the fact that this map transforms one Hilbert basis of the space Z into another. The canonicity results from identities

$$\langle iU_a u, U_a v\rangle_Z = \langle U_a iu, U_a v\rangle_Z = \langle iu, v\rangle_Z$$

(we use the orthogonality of U_a).

The estimate (3.6) for the spectrum of the Sturm–Liouville problem is well-known [PT,Ma]. To prove (3.7) let us observe that for the eigen-functions $\varphi_j(x;a)$ one has the estimate

$$\|\varphi_j(a_1)-\varphi_j(a_2)\|_0 \le C\sup_x |V(x;a_1)-V(x;a_2)|/j \le C_1|a_1-a_2|/j \tag{3.9}$$

(see [PT,Ma]). As

$$\frac{\partial^2}{\partial x^2}\varphi_j(x;a) = \big(V(x;a)-\lambda_j(a)\big)\varphi_j(x;a)\,,$$

then we get from estimates (3.6), (3.9) that $\|\varphi_j^{\pm}(a_1)-\varphi_j^{\pm}(a_2)\|_2 \le C_2|a_1-a_2|j$. From (3.9), the last inequalities and the interpolation inequality (see Appendix A to Part 3 below and [RS]) it follows that for all $s\in[0,2]$

$$\|\varphi_j^{\pm}(a_1)-\varphi_j^{\pm}(a_2)\|_s \le C_2 j^{s-1}|a_1-a_2|\,. \tag{3.10}$$

Let $u\in Z_s$ and

$$u=\sum u_k^{\pm}\varphi_k^{\pm}(x)\,,\quad \|u\|_s^2=\sum |u_k^{\pm}|^2 k^{2s}<\infty$$

(we use the equality $\|\varphi_j^{\pm}\|_s = j^s$). Then

$$\begin{aligned}\|U_{a_1}u-U_{a_2}u\|_s &= \Big\|\sum_k u_k^{\pm}(U_{a_1}-U_{a_2})\varphi_k^{\pm}(x)\Big\|_s \le\\ &\le \sum_k |u_k^{\pm}|\,\|\varphi_k^{\pm}(x;a_1)-\varphi_k^{\pm}(x;a_2)\|_s \le\\ &\le C_s'\Big(\sum_k |u_k^{\pm}|^2k^{2s}\Big)^{1/2}|a_1-a_2|\Big(\sum k^{-2}\Big)^{1/2}\le\\ &\le C_s''|a_1-a_2|\,\|u\|_s\,,\end{aligned}$$

and we get the estimate (3.7). The estimate (3.8) results from the inequality

$$\|\varphi_j^{\pm}(x;a)-\varphi_j^{\pm}(x)\|_s \le C_j^1 j^{s-1}$$

in the same way as (3.7) results from (3.10). □

By Lemma 3.2 and Theorem 2.2 from Part 1, the substitution $u=U_a v$ transforms solutions of equation (3.5) to the solutions of the equation

$$\dot v = J\big(A(a)v+\varepsilon\nabla H(v;a)\big) \tag{3.11}$$

with the hamiltonian $\frac12\langle A(a)v,v\rangle + H(v;a)$, where

$$A(a)=U_a^*A_aU_a\,,\quad H=H_0(U_av;a)\,.$$

So

$$\nabla H(v;a) = U_a^*\nabla H_0(U_av;a)$$

and

$$A(a)\varphi_j^{\pm}(a) = \lambda_j^A(a)\varphi_j^{\pm} \ \forall j .$$

The operators $A(a)$ commute and for the transformed equation (3.11) the relations (1.2), (1.3) trivially hold.

Equation (3.5) with $\varepsilon = 0$ is the linear Schrödinger equation

$$\dot{u} = i(-u_{xx} + V(x;a)u) , \quad u(t) \in (\mathring{H}^1 \cap H^2)(0,\pi;\mathbb{C}) ,$$

and it has invariant n-tori of the form

$$T_a^n(I) = \Big\{ \sum_{j=1}^{n} (\alpha_j^+ + i\alpha_j^-)\varphi_j(x;a) \mid \alpha_j^{+^2} + \alpha_j^{-^2} = 2I_j > 0 \Big\} . \tag{3.12}$$

Let a Borel set $\mathcal{J} \subset \mathbb{R}_+^n$ be as in Part 2.1 and $\mathcal{T}_a = \cup\{T_a^n(I) \mid I \in \mathcal{J}\}$. For every $a \in \mathfrak{A}$, $U_a^{-1}(T_a^n(I))$ is an invariant torus $T(I)$ of equation (3.11) with $\varepsilon = 0$. This torus does not depend on a and the tori $T(I)$ with $I \in \mathcal{J}$ form a surface $\mathcal{T}$, $\mathcal{T} = \cup\{T(I)\} = U_a^{-1}\mathcal{T}_a$. This surface is contained in the domains $U_a^{-1}B^c$ (see Lemma 3.1) with some neighborhood O^c, independent of a:

$$\mathcal{T} \subset O^c \subset \cap U_a^{-1} B^c .$$

Let us check that Theorem 1.1 with

$$d_J = 0,\ d_A = 2,\ d_H = 0,\ d = 2 ,$$

may be applied to the equation (3.11). Indeed, the validity of assumption 1) with O^c as above results from Lemma 3.1, (3.7), (3.8); assumption 2) with $d_1 = d_A = 2$ results from (3.6) and from the well-known asymptotic $\lambda_j = j^2 + O(1)$ (see [PT], [Ma]). So by applying Corollary 1.1 we get the following statement:

Theorem 3.3. Let $\rho < 1$ and a_0 be a point in $\mathfrak{A}$ such that

$$\left|\det\big(\partial\lambda_j^A(a_0)/\partial a_k\big) \mid 1 \le j,k \le n\right| \ge K_0 > 0 . \tag{3.13}$$

Then there exist large enough integers j_1, M_1, such that if

$$\left|\lambda_1^A(a_0)s_1 + \lambda_2^A(a_0)s_2 + \cdots + \lambda_{j_1}^A(a_0)s_{j_1}\right| \ge K_3 > 0 \tag{3.14}$$

for all integer j_1-vectors s such that

$$|s| \le M_1, \quad 1 \le |s_{n+1}| + \cdots + |s_{j_1}| \le 2 ,$$

then there exists $\delta_a > 0$ and for sufficiently small $\varepsilon > 0$ there exists a Borel subset Θ_ε of the set $O(a_0, \delta_a, \mathfrak{A}) \times \mathcal{J}$ and analytic embeddings

$$\Sigma_{(a,I)}^\varepsilon : \mathbb{T}^n \longrightarrow (\mathring{H}^1 \cap H^2)(0,\pi;\mathbb{C}) , \quad (a,I) \in \Theta_\varepsilon ,$$

with the following properties:

a) $\operatorname{mes}\Theta_\varepsilon[I] \to \operatorname{mes} O(a_0, \delta_a, \mathfrak{A})$ when $\varepsilon \to 0$, uniformly with respect to I;

b) every torus $\Sigma^\varepsilon_{(a,I)}(\mathrm{T}^n)$ is invariant for the equation (3.5) and is filled with weak in $(\mathring{H}^1 \cap H^2)$ time-quasiperiodic solutions of (3.5) of the form $t \longmapsto \Sigma^\varepsilon_{(a,I)}(q_0 + \omega' t)$
(q_0 is an arbitrary point from T^n and $\omega' = \omega'(a, I, \varepsilon) \in \mathrm{R}^n$);

c) $\operatorname{dist}_{H^2}(\Sigma^\varepsilon_{(a,I)}(\mathrm{T}^n), T^n_a(I)) \le \varepsilon^\rho$ and $|\omega - \omega'| \le \varepsilon^\rho$;

d) the numbers j_1, M_1 depend on K_0, K_1 and n only; the radius δ_a and the rate of the convergence in a) depends on the same numbers and K_3.

Let us discuss assumptions (3.13), (3.14) of the theorem. To do it we consider the mapping $\mathcal{U}$ from the set $\mathfrak{A}$ into the space $C[0,\pi]$ of potentials $V(x)$,

$$\mathcal{U} : \mathfrak{A} \longrightarrow C[0,\pi]\,, \quad a \longmapsto V(\cdot; a)\,.$$

Every eigenvalue λ^A_j is an analytic function of the potential $V(x)$. So condition (3.14) means that the point $\mathcal{U}(a_0)$ lies in the space $C[0,\pi]$ outside the zero set of some nontrivial analytic function. To discuss assumption (3.12) let us use the formula

$$\frac{\partial \lambda^A_j}{\partial a_k}(a_0) = \int_0^\pi \varphi_j^2(x; a_0) \frac{\partial V(x; a_0)}{\partial a_k}\, dx \tag{3.15}$$

(see [PT,RS]). It is proven in [PT] that the system of the functions $\{\varphi_1^2(\cdot; a), \dots, \varphi_n^2(\cdot; a)\}$ is linearly independent for all a. Thus the function

$$(\xi_1(x), \dots, \xi_n(x)) \longmapsto \det\Big\{ \int \varphi_j^2(x; a)\xi_\ell(x) dx \;\Big|\; 1 \le j, \ell \le n \Big\}$$

turns out to be a non-trivial n-form on the space $C[0,\pi]$ and the condition (3.13) means that the restriction of this n-form on the n-dimensional image of the tangent mapping

$$\mathcal{U}_*(a_0) : \mathrm{R}^n \longrightarrow T_{\mathcal{U}(a_0)} C[0,\pi] \simeq C[0,\pi]$$

is nondegenerate.

So the assumption (3.13) + (3.14) is a non-degeneracy assumption on the 1-jet of the map $\mathcal{U}$ at the point a_0.

Remark. Theorem 1.1 is also applicable to study equation (3.5) under the Neumann boundary conditions. That is equivalent to study it in the space of even periodic with respect to x functions,

$$x \in \mathrm{R}\,, \quad u(t, x) \equiv u(t, x + 2\pi)\,, \quad u(t, x) \equiv u(t, -x)\,, \tag{3.16}$$

if the functions V and χ are even periodic and smooth in x. Now one has to take for spaces $\{Z_s\}$ the spaces of even periodic Sobolev functions. In such a case the formula (3.4) defines an analytic mapping from the space Z_s into itself for every $s \ge 1$. So Theorem 1.1 is applicable with arbitrary $d \ge 1$ and in the case of the problem (3.5), (3.16) one may prove existence of arbitrary smooth invariant tori (i.e. being in the space $H^k(0, \pi; \mathrm{C})$ with k arbitrarily large) at a distance of order ε from T_a, provided that $\varepsilon \le \varepsilon(k)$ for some positive $\varepsilon(k)$.

4. Schrödinger equation with random potential

We consider nonlinear Schrödinger equation with real random potential $V_\omega(x)$ under the Dirichlet boundary conditions

$$\begin{aligned} \dot{u} &= i\left(-u_{xx} + V_\omega(x)u + \varepsilon \frac{\partial}{\partial |u|^2}\chi(x,|u|^2)u\right), \\ u &= u(t,x), \quad u(t,0) \equiv u(t,2\pi) \equiv 0 . \end{aligned} \tag{4.1}$$

The function χ is supposed to satisfy relation (3.2) (with a one-point parameter set $\mathfrak{A}$).

Let us denote by QP_ε the random subset of the phase-space Z, equal to the union in Z of all the curves corresponding to time-quasiperiodic solutions of (4.1) with zero Lyapunov exponents. Our goal is to prove that the set QP_ε is asymptotically dense in the phase space when $\varepsilon \to 0$: for every $\mathfrak{z}(x) \in Z_2$

$$\begin{gathered} \operatorname{dist}_2\left(\mathfrak{z}(x), QP_\varepsilon\right) \longrightarrow 0 \quad (\varepsilon \to 0) \\ \text{in probability.} \end{gathered} \tag{4.2}$$

The relation (4.2) can be proven for random potentials $V_\omega(x)$ "with good randomness properties". In order to avoid technical difficulties, unconnected with the main contents of the book, we restrict ourself to potentials V, given by a centered, smooth enough, even periodic random process:

$$V_\omega(x) = \sum_{j=1}^{\infty} \omega_j \kappa_j \cos jx, \quad K^{-1} j^{-2} \le \kappa_j \le K j^{-2} . \tag{4.3}$$

Here $\omega_1, \omega_2, \ldots$ are independent random variables, uniformly distributed in the segment $\Delta = \left[-\frac{1}{2}, \frac{1}{2}\right]$. So V_ω is a random process with the probability space $\Omega = \Delta \times \Delta \times \ldots$, endowed with the σ-algebra $\mathcal{B}$ of cylindric sets and usual Kolmogorov measure $d\mu = d\omega_1 \otimes d\omega_2 \otimes \cdots$ on them. We also endow Ω with the metric dist, where

$$\operatorname{dist}(\omega, \tilde{\omega}) = \sum_{j=1}^{\infty} j^{-2} |\omega_j - \tilde{\omega}_j| .$$

The set Ω is compact in the corresponding topology and $\mathcal{B}$ is the σ-algebra of the Borel subsets of Ω. Below we use no random process theory but elementary properties of the measure $d\mu$ only. Everything we need can be found e.g., in [SG].

Theorem 4.1. The convergence (4.2) holds for each point $\mathfrak{z} \in Z_2$.

Proof. We use the notations from the previous part. So $\{\varphi_j(x;\omega)\}$ is the eigenbasis of the operator $-\partial^2/\partial x^2 + V_\omega$ and $\{\lambda_j^A(\omega)\}$ is the spectrum of this operator.

Let us fix any $\delta > 0$.

Lemma 4.2. There exists a non-random integer n such that $\|\mathfrak{z} - \mathfrak{z}_0\|_2 < \delta/3$ for a vector $\mathfrak{z}_0$ equal to the sum of the first $2n$ terms of the decomposition $\mathfrak{z}$ in the basis $\{\varphi_j^{\pm}(\omega)\}$,

$$\mathfrak{z}_0 = \sum_{j=1}^{n} \tilde{z}_j^{\pm}(\omega)\varphi_j^{\pm}(\omega) . \tag{4.4}$$

Proof. For $\omega = 0$ the vectors $\varphi_j^{\pm}(0)$ are equal to $\varphi_j^{+}(0) = \varphi_j(x)$, $\varphi_j^{-}(0) = i\varphi_j(x)$ with

$$\varphi_j(x) = \sin \frac{1}{2} jx .$$

We can write $\mathfrak{z} \in Z_2$ as $\mathfrak{z} = \sum z_j^{\pm}\varphi_j^{\pm}(0)$, where

$$\sum (|z_j^{+}|^2 + |z_j^{-}|^2) j^4 = K^2 < \infty . \tag{4.5}$$

So there exists $n' < \infty$ such that

$$\left\| \sum_{j=n'+1}^{\infty} z_j^{\pm}\varphi_j^{\pm}(0) \right\|_2 \leq \delta/C_* \tag{4.6}$$

(the constant C_* will be chosen later).

Let us fix some $\theta \in (0,1/2)$. For each ω the potential V_ω belongs to the Sobolev space $H^\theta(0,\pi)$. In the same way as in Lemma 3.2 (see estimate (3.10)), we find that

$$\|\varphi_j^{\pm}(\omega) - \varphi_j^{\pm}(0)\|_{2+\theta} \leq C j^{1+\theta} . \tag{4.7}$$

Let us write $\mathfrak{z}$ in the form

$$\mathfrak{z} = \mathfrak{z}^1 + \mathfrak{z}^2 , \quad \mathfrak{z}^1 = \sum z_j^{\pm}\varphi_j^{\pm}(\omega) , \quad \mathfrak{z}^2 = \sum z_j^{\pm}(\varphi_j^{\pm}(0) - \varphi_j^{\pm}(\omega)) .$$

By (4.5), (4.7) we can estimate the norm of the vector $\mathfrak{z}^2$ as follows:

$$\begin{aligned} \|\mathfrak{z}^2\|_{2+\theta} &\leq \sum |z_j^{\pm}| \, \|\varphi_j^{\pm}(\omega) - \varphi_j^{\pm}(0)\|_{2+\theta} \\ &\leq C\sum |z_j^{\pm}| j^{1+\theta} \leq C(\sum |z_j^{\pm}|^2 j^4)^{1/2}(\sum j^{-2(1-\theta)})^{1/2} \leq C_1 K . \end{aligned}$$

So $\mathfrak{z}^2 = \sum y_j^{\pm}(\omega)\varphi_j^{\pm}(\omega)$, where $\sum |y_j^{\pm}(\omega)|^2 j^{4+2\theta} \leq CK^2$. Thus

$$\left\| \sum_{n+1}^{\infty} y_j^{\pm}(\omega)\varphi_j^{\pm}(\omega) \right\|_2^2 \leq C \sum_{n+1}^{\infty} |y_j^{\pm}(\omega)|^2 j^4 \leq n^{-2\theta} C_1 K^2 . \tag{4.8}$$

We also see from the estimate for the norm of the vector $\mathfrak{z}^2$ that the norm of the linear operator

$$Z_2 \longrightarrow Z_2 , \quad \varphi_j(0) \longmapsto \varphi_j(\omega) \quad \forall j$$

is majorized by $\tilde{C} = 1 + C_1$. So due to the estimate (4.6),

$$\left\| \sum_{n+1}^{\infty} z_j^{\pm} \varphi_j^{\pm}(\omega) \right\|_2 \leq \tilde{C}\delta / C_* . \tag{4.9}$$

Let choose C_* in (4.6) and $n \geq n'$ in such a way that $\tilde{C}/C_* < \delta/6, n^{-2\theta} C_1 K^2 < (\delta/6)^2$, and denote

$$\mathfrak{z}_0 = \sum_{j=1}^{n} (z_j^{\pm} + y_j^{\pm}(\omega)) \varphi_j^{\pm}(\omega) .$$

By the estimates (4.8) and (4.9), $\|\mathfrak{z} - \mathfrak{z}_0\|_2 < \delta/3$, and the lemma is proven with $\tilde{z}_j^{\pm}(\omega) = z_j^{\pm} + y_j^{\pm}(\omega)$ for $j = 1, \ldots, n$. □

The vector $\mathfrak{z}_0$ lies in a torus T_ω^n of the form (3.12), invariant for the flow of the linear equation. We can suppose that the torus T_ω^n is "essentially n-dimensional":

$$\left|x_j^{+}\right|^2 + \left|y_j^{+}\right|^2 > (\delta/3n)^2 \quad \forall j . \tag{4.10}$$

Otherwise we can neglect the terms violating (4.10) and obtain a vector of the form (4.4) with smaller n, which approximates $\mathfrak{z}$ with the accuracy $2\delta/3$.

Now we are going to approximate the vector $\mathfrak{z}_0$ by the quasiperiodic solutions. To do it, we split a sequence $\omega = (\omega_1, \omega_2, \ldots)$ to $\omega = (\omega^n, \omega^{\infty - n})$, where $\omega^n = (\omega_1, \ldots, \omega_n)$ and $\omega^{\infty - n} = (\omega_{n+1}, \omega_{n+2}, \ldots)$. Correspondingly, we split Ω and $d\mu$ to the products of the probability spaces and the measures:

$$\Omega = \Omega^n \times \Omega^{\infty - n} , \quad d\mu = d\mu^n \otimes d\mu^{\infty - n} .$$

The function

$$D(\omega) = \det \left\{ \frac{\partial \lambda_j^A(\omega)}{\partial \omega_k} \mid 1 \leq j, k \leq n \right\}$$

is continuous in Ω because λ_j^A is an analytic function of the potential $V(x) \in L_\infty(0, 2\pi)$ (see [Ma,PT]). Thus, the set $\mathcal{E}_0 = D^{-1}(0)$ is closed.

Lemma 4.3 $\mu(\mathcal{E}_0) = 0$.

Proof. Due to the formula (3.15) with $\varphi_j(x)$ as in the proof of Lemma 4.2,

$$\begin{aligned} \frac{\partial \lambda_j^A}{\partial \omega_k}(0) &= \text{const}\, \kappa_j \int \cos kx \sin^2 \frac{1}{2} jx \, dx \\ &= \text{const}\, \kappa_j \int \cos kx \cos jx \, dx = \text{const}\, \kappa_j \delta_{jk} . \end{aligned}$$

Hence, $D(0) \neq 0$.

As the eigenvalues $\{\lambda_j^A\}$ C^1-smooth depend on the potential $V(x) \in L^\infty(0, 2\pi)$ and

$$\left| V_\omega(x) - \sum_{j=1}^{M} \omega_j \kappa_j \cos jx \right| \leq C M^{-1} ,$$

then $|D(0,\omega^{\infty-M})| \geq \frac{1}{2}|D(0)|$, if M is large enough. The function

$$\omega^M \longmapsto D(\omega^M,\omega^{\infty-M})$$

is analytic and is not identically zero. Thus its zero-set has zero measure; so for every $\omega^{\infty-M}$

$$\mu^M\{\omega^M \mid (\omega^M,\omega^{\infty-M}) \in \mathcal{E}_0\} = 0\,.$$

Now the lemma's assertion results from Fubini theorem. □

For $t > 0$ let us define the open sets $\mathcal{E}_t$,

$$\mathcal{E}_t = \{\omega \mid |D(\omega)| < t\}\,.$$

These sets form a decreasing sequence of open domains and $\bigcap_{t>0} \mathcal{E}_t = \mathcal{E}_0$. So

$$\lim_{t\to 0} \mu(\mathcal{E}_t) = \mu(\mathcal{E}_0) = 0$$

and there exists $K_0 > 0$ such that

$$\mu(\mathcal{E}_{K_0}) < \frac{1}{3}\gamma\,.$$

Now we are going to apply Theorem 3.3 with fixed $\omega^{\infty-n}$ and $a = \omega^n \in \Omega^n$, in a neighborhood of a point $a_0 = \omega_0^n$ such that $(\omega_0^n,\omega^{\infty-n}) \notin \mathcal{E}_{K_0}$. The assumptions (3.1), (3.2) and (3.13) are trivially fulfilled. So the theorem can be applied provided that the assumption $|R^\ell(\omega)| \geq K_3$ holds with some $K_3 > 0$ for a finite number G of resonance relations R^ℓ of the form

$$R^\ell(\omega) = \sum_{j=1}^{j_1} \ell_j \lambda_j^A(\omega^n,\omega^{\infty-n})\,, \quad \ell \in \mathcal{L} \subset Z^{j_1}\,, \quad |\mathcal{L}| = G\,.$$

(The number j_1 and the set $\mathcal{L}$ are independent from $(\omega_0^n,\omega^{\infty-n})$ due to statement d) of the theorem.) Each function $R^\ell(\omega)$ is not identically zero because the image of the map $\omega \longmapsto (\lambda_1^A,\lambda_2^A,\ldots)$ forms an open subset in the space of sequences (Borg theorem, see [PT],p. 55). Thus, as in Lemma 4.3, $\mu((R^\ell)^{-1}(0)) = 0$ and there exists $K_3 > 0$ such that $\mu(\mathcal{E}^{K_3}) < \gamma/3$, where for $t > 0$

$$\mathcal{E}^t = \{\omega \mid |R^\ell(\omega)| < 2t \quad \forall \ell \in \mathcal{L}\}\,.$$

Let us set $\mathcal{E} = \mathcal{E}_{K_0} \cup \mathcal{E}^{K_3}$. Then $\mu(\mathcal{E}) < \frac{2}{3}\gamma$. The set $\Omega\backslash\mathcal{E}$ is a closed subset of compact set Ω; so it is compact.

Given a point $(\omega_0^n,\omega_0^{\infty-n}) \in \Omega\backslash\mathcal{E}$, we can find a neighbourhood $U^{\infty-n}$ of $\omega_0^{\infty-n}$ in $\Omega^{\infty-n}$ such that $|R^\ell(\omega_0^n,\omega^{\infty-n})| \geq K_3$ for all $\ell \in \mathcal{L}$ and every $\omega^{\infty-n} \in U^{\infty-n}$. We can apply Theorem 3.3 to equation (4.1) with fixed $\omega^{\infty-n} \in U^{\infty-n}$, treating $\omega^n \in \Omega^n$ as a parameter a with $a_0 = \omega_0^n$. So for fixed $\omega^{\infty-n}$ we can construct a ball $U^n = O(\omega_0^n, \delta_a, \Omega^n)$ and for each $t > 0$ construct a measurable subset $U_\varepsilon^n(t) = U_\varepsilon^n(\omega_0^n, \omega^{\infty-n}; t)$ of U^n such that

$$\operatorname{mes} U_\varepsilon^n(t) < t \tag{4.11}$$

for $\varepsilon < \varepsilon(t)$. For $\omega^n \in (U^n \backslash U_\varepsilon^n(t))$ the equation (4.1) has invariant n-torus $T_{\omega,\varepsilon}^n$, which is $\sqrt{\varepsilon}$-close to the torus T_ω^n and is filled with time-quasiperiodic solutions. The function $t \longmapsto \varepsilon(t)$ is independent from $(\omega_0^n, \omega^{\infty-n})$ by the assertion d) of the theorem.

The sets $U^n \times U^{\infty-n}$, corresponding to the points $(\omega_0^n, \omega_0^{\infty-n})$, form an open covering of the compact set $\Omega \backslash \mathcal{E}$. Let us fix some finite subcovering $\{U_j^n \times U_j^{\infty-n} \mid j = 1, \dots, M\}$, where the set $U_j^n \times U_j^{\infty-n}$ corresponds to a point $(\omega_0^n, \omega_0^{\infty-n}) = (\omega_j^n, \omega_j^{\infty-n})$, $j = 1, \dots, M$. We choose in (4.11) $t = \gamma/(3M)$ and define the sets $D_{j\varepsilon} \subset U_j^n \times U_j^{\infty-n}$ equal to

$$D_{j\varepsilon} = \left\{ (\omega^n, \omega^{\infty-n}) \in U_j^n \times U_j^{\infty-n} \mid \omega^n \in U_{j\varepsilon}^n(\omega_j^n, \omega^{\infty-n}; t) \right\}.^{10)}$$

By Fubini theorem and (4.11) with $t = \gamma/(3M)$

$$\mu(D_{j\varepsilon}) = \int\limits_{U_j^{\infty-n}} d\mu^{\infty-n} \Big(\mu^n (U_{j\varepsilon}^n(\omega_j^n, \omega^{\infty-n}; t)) \Big) \leq \frac{\gamma}{3M} \mu^{\infty-n}(U_j^{\infty-n}) \leq \frac{\gamma}{3M}.$$

Let us set

$$\mathcal{M}_\varepsilon = \mathcal{E} \cup \Big(\bigcup_{j=1}^{M} D_{j\varepsilon} \Big).$$

Then

$$\mu(\mathcal{M}_\varepsilon) \leq \mu(\mathcal{E}) + \sum \mu(D_{j\varepsilon}) \leq \gamma.$$

The set $\Omega \backslash \mathcal{M}_\varepsilon$ is contained in the union

$$\bigcup_{j=1}^{M} (U_j^n \times U_j^{\infty-n} \backslash D_{j\varepsilon}).$$

So, for $\omega \notin \mathcal{M}_\varepsilon$ the equation (4.1) has a quasiperiodic solution at a distance $\leq \sqrt{\varepsilon}$ from the point $\mathfrak{z}_0$. If $\varepsilon < \delta^2/9$, then this solution is δ-close to the point $\mathfrak{z}$. So

$$\mu\big\{ \operatorname{dist}_2(\mathfrak{z}, QP_\varepsilon) > \delta \big\} < \gamma$$

provided that $\varepsilon < \varepsilon(\delta, \gamma)$. We can choose $\delta, \gamma > 0$ as small as we wish; thus the convergence (4.2) is proven. □

---

[10] This set is measurable due to given below in Part 3 refined version of Theorem 1.1, where the hamiltonian $\mathcal{H}$ of equation (1.5) is allowed to depend on an additional parameter from arbitrary metric space (one should take for such a parameter the vector $\omega^{\infty-n} \in U_j^{\infty-n}$).

Remark 1. The random process (4.3) is the minimum one for which the given above proof works. We can replace it, for example, by the smooth periodic process

$$W_{\omega,\bar{\omega}}(x) = \Big(\sum_{j=1}^{\infty} \omega_j \kappa_j \cos jx\Big) + \Big(\bar{\omega}_0 \kappa_0 + \sum_{j=1}^{\infty} \bar{\omega}_j \bar{\kappa}_j \sin jx\Big) = V_\omega(x) + \bar{V}_{\bar{\omega}}(x)$$

(the random variables $\omega_1, \omega_2, \ldots \bar{\omega}_0, \bar{\omega}_1, \bar{\omega}_2, \ldots$ are independently and uniformly distributed in the segment $[-\frac{1}{2}, \frac{1}{2}]$). One can fix $(\bar{\omega}_0, \bar{\omega}_1, \ldots)$, repeat the proof for potential $V_\omega(x)$ replaced by $V_\omega(x) + \bar{V}_{\bar{\omega}}(x)$ and obtain the convergence (4.2) for equation (4.1) with the potential $W_{\omega,\bar{\omega}}(x)$.

Remark 2. The convergence (4.2) is a general property of Hamiltonian PDE's with one-dimensional x variables and random coefficients. In particular, the same proof imply the convergence (4.2) for the nonlinear string equation (see below equation (6.1)) with a random potential.

Remark 3. If nonlinear term χ in equation (4.1) or (6.1) with random potential $V_\omega(x)$ is "nondegenerate" (e.g., if in (6.1) $\chi = w^3$), then the scheme described in the part 3.2.B of Introduction can be applied to study the equations. For a fixed vector $\mathfrak{z}$ we can choose the approximating "n-dimensional" vector $\mathfrak{z}_0$ as in Lemma 4.2 and put the hamiltonian of the equation to the normal form (28) (see Introduction) in a neighbourhood of the corresponding torus T_ω^n, by means of a transformation, which is well-defined for almost all ω. After this in a nondegenerate situation we can use the amplitude-frequency modulation (27) to prove the existence of quasiperiodic solutions near $\mathfrak{z}_0$, without extra restrictions on the admissible set of parameters ω. Thus, for nondegenerate nonlinearities the convergence (4.2) holds for almost all ω (not only in probability).

5. Nonlinear Schrödinger equation on the real line

In §§3,4 we have applied Theorem 1.1 to construct time-quasiperiodic solutions of nonlinear Schrödinger equation under the Dirichlet boundary conditions

$$0 \le x \le \pi\,, \quad u(t,0) \equiv u(t,\pi) \equiv 0\,.$$

This equation describes a one-dimensional quantum particle locked at the potential well $[0,\pi]$ by means of a potential equal to infinity out of the well. We shall see now (without going to details) that a similar phenomenon occurs for potentials, which grow at infinity fast enough.

We start with perturbing the quantized harmonic oscillator:

$$\dot{u} = i\big(-u_{xx} + (x^2 + V_0(x;a))u + \varepsilon \nabla H(u;a)\big)\,, \tag{5.1}$$

$$u = u(t,x)\,, \quad x \in \mathbb{R}\,,$$

where V_0 is a smooth function, analytic in a, vanishing at $x = \infty$. The equation (5.1) can be treated in the same way as the nonlinear Schrödinger equation under the Dirichlet boundary conditions in §3. The spaces Z_s should be chosen equal to the domains of definition of the operators $(-\partial^2/\partial x^2 + x^2)^s$. The operators $\mathcal{A}(a) = -\partial^2/\partial x^2 + x^2 + V_0$ and J $(Ju = iu)$ define morphisms of this scale of order one and zero respectively. After an unessential linear transformation of the phase space, similar to the transformation U_a from §3, the relations (1.2), (1.3) hold with $\lambda_j^J(a) \equiv 1$ and $\{\lambda_j^A(a)\}$ equal to the spectrum of the operator $\mathcal{A}(a)$. The first assumption of Theorem 1.1 holds with $d_J = 0$, $d_A = 1$, provided that the functional H_a defines an analytic gradiental map $\nabla H_a : Z_d \to Z_{d-d_H}$ of some order $d_H < 0$. The eigenvalues $\{\lambda_j(a) = \lambda_j^A(a)\}$ of the operator $\mathcal{A}(a)$ obey Bohr's quantisation law: $\lambda_n \sim C(n+1/2)$. Moreover, $|\lambda_n - C(n+1/2)| \le C_1 n^{-1/2}$ (see [HR, BS]). Hence the second assumption of the theorem holds with $d_1 = 1$, $r = 2$ and $d_{1,1} = 0$, $d_{1,2} = -1/2$.

Thus, Theorems 1.1, 1.2 can be applied to equation (5.1) with typical potentials V_0, if the gradiental map ∇H is of a negative order. In particular, if

$$H = \frac{1}{2}\int \chi\left(|u * \xi(x)|^2 ; a\right) dx$$

($u * \xi$ is the convolution with a smooth real-valued function $\xi(x)$, decreasing at infinity), then

$$\nabla H(u; a) = \xi * \left[\chi'\left(|u * \xi(x)|^2 ; a\right) u * \xi(x)\right] .$$

So $d_H = -1$ provided that $\chi(p; a) = O\left(|p|^2\right)$ as $p \to 0$.

We can also consider a perturbed unharmonic oscillator:

$$\dot{u} = i\Big(-u_{xx} + \left(x^2 + \mu x^4 + V_0(x; a)\right) u + \varepsilon \chi'\left(|u|^2 ; a\right) u\Big) , \tag{5.2}$$

where $\mu > 0$, potential V_0 is as above and a smooth real-valued function χ is analytic in a. Now

$$\lambda_n(a) = C_1(n+1/2)^{4/3} + C_2(n+1/2)^{2/3} + O(1) , \tag{5.3}$$

see [HR]. So the assumption 2) of Theorem 1 holds with $d_1 = 4/3$, $d_J = 0$, $r = 4$ and $d_{1,j} = (4-j)/3$ for $j = 1, \ldots, r$.

The spaces Z_s now should be chosen equal to the domains of definition of the operators $A^{(3/4)s}$. Then the operators $\mathcal{A}(a)$ and J define morphisms of the scale of order 4/3 and zero; the gradiental map

$$Z_{4/3} \longrightarrow Z_{4/3} , \quad u \longmapsto \chi'\left(|u|^2 ; a\right) u$$

is analytic. So assumption 1) holds with $d_1 = 4/3$, $d_J = d_H = 0$.

Thus, Theorems 1.1, 2.2 can be applied to equation (5.2). In particular, assertions of Theorem 3.3 hold for this equation, provided that the nondegeneracy and nonresonance relations (3.13), (3.14) are fulfilled. So typically equation (5.2) has many time-quasiperiodic solutions, if ε is small enough.

In particular, if $\varphi_n(x; a)$ is a stationary state and $\exp\big(i\lambda_n(a)t\big)\varphi_n(x; a)$ is the corresponding solution of linear equation $(5.2)|_{\varepsilon=0}$, then this solution persists as a time-periodic solution of the perturbed equation (5.2) for most one-dimensional parameters a, if $\lambda_n'(a) \not\equiv 0$ and

$$s\lambda_n(a) \not\equiv 2\lambda_j(a) , \quad s\lambda_n(a) \not\equiv \lambda_j(a) \pm \lambda_k(a) ,$$

where s is an arbitrary integer and the numbers n, j, k are pairwise different.

6. Nonlinear string equation

The next application of our theorem will be to the equation of oscillation of a string with fixed ends in a nonlinear-elastic medium depending on a parameter $a \in \mathfrak{A}$:

$$\frac{\partial^2}{\partial t^2} w = (\partial^2/\partial x^2 - V(x;a))w - \varepsilon \frac{\partial}{\partial w}\chi(x,w;a)\,; \tag{6.1}$$

$$w = w(t,x)\,, \quad 0 \le t \le \pi\,; \quad w(t,0) \equiv w(t,\pi) \equiv 0\,. \tag{6.2}$$

To write this non-linear boundary value problem in the form (1.5), we need some preliminary work. Let us suppose that $V : [0,\pi] \times \tilde{\mathfrak{A}} \longrightarrow \mathbb{R}_+$ is a smooth function. The differential operator $-\partial^2/\partial x^2 + V(x;a)$ defines a positive selfadjoint operator $\mathcal{A}_a$ in the space $L_2(0,\pi;\mathbb{R})$ with the domain of definition $(\dot{H}^1 \cap H^2)(0,\pi;\mathbb{R})$. The space $\mathcal{Z} = D(\sqrt{\mathcal{A}_a})$ is the Sobolev space $\dot{H}^1(0,\pi;\mathbb{R})$ with the scalar product

$$\langle u,v\rangle^{(a)} = \int_0^\pi (u_x v_x + V(x;a)uv)\,dx\,.$$

For $t \ge 0$, let $\mathcal{Z}_t$ be the space $\mathcal{Z}_t = D(\mathcal{A}_a^{(t+1)/2})$ with the norm $\|u\|_t^{(a)} = \left\|\mathcal{A}_a^{t/2}(u)\right\|_0^{(a)}$. In particular, $\|u\|_0^{(a)} = \left(\langle u,u\rangle^{(a)}\right)^{1/2}$. For $-t \le 0$ let $\mathcal{Z}_{-t}$ be the space dual to $\mathcal{Z}_t$ with respect to the scalar product $\langle\cdot;\cdot\rangle^{(a)}$ (see Part 1.1). Let us set $Z_t^{(a)} = \mathcal{Z}_t \times \mathcal{Z}_t$ with the natural norm and scalar product, the last will be also denoted as $\langle\cdot,\cdot\rangle^{(a)}$. In the scale $\{Z_t^{(a)}\}$ let us consider the operator J_a of order $d_J = 1$,

$$J_a : Z_t^{(a)} \longrightarrow Z_{t-1}^{(a)}\,, \quad w = (w_1,w_2) \longmapsto \left(\mathcal{A}_a^{1/2} w_2, -\mathcal{A}_a^{1/2} w_1\right).$$

This operator is antiselfadjoint in $Z^{(a)} = Z_0^{(a)}$ with the domain of definition $D(J_a) = Z_1^{(a)}$. The triple

$$\left\{Z^{(a)}, \{Z_s^{(a)} \mid s \in \mathbb{R}\}\,, \langle \bar{J}_a dw, dw\rangle^{(a)}\right\},$$

where $\bar{J}_a = -(J_a)^{-1}$, is a symplectic Hilbert scale, depending on the parameter a.

As in §3, we denote by $\{\varphi_j(x;a) | j \ge 1\}$ the complete system of the eigenfunctions of the operator $-\partial^2/\partial x^2 + V(x;a)$ and denote by $\{\lambda_j^{(a)}\}$ the corresponding eigenvalues. We set

$$\varphi_j^{+(a)} = (\varphi_j(x;a),0)\left(\lambda_j^{(a)}\right)^{-1/2}, \quad \varphi_j^{-(a)} = (0,\varphi_j(x;a))\left(\lambda_j^{(a)}\right)^{-1/2}.$$

The functions $\left\{(\lambda_j^{(a)})^{-s/2}\varphi_j^{\pm(a)} \mid j \ge 1\right\}$ forms a Hilbert basis of each space $Z_s^{(a)}$, $s \in \mathbb{R}$, and the relations (1.2) hold with $\lambda_j^J = \sqrt{\lambda_j^{(a)}}$ and $\varphi_j^{\pm} = \varphi_j^{\pm(a)}$.

Let the function $\chi(x,w;a)$ and the domain $O^c \subset \mathbb{C}$ be the same as in §4 with the additional property

$$\chi(0,0;a) \equiv \chi(\pi,0;a) \equiv 0\,.$$

We define the functional H^0,

$$H^0(w_1, w_2; a) = \int_0^\pi \chi(x, w_1(x); a)\, dx\ . \tag{6.3}$$

Lemma 6.1. For any $R > 0$ there exists a complex δ-neighborhood B^c of the ball $\{u \in Z_1^{(a)} \mid \|u\|_1^{(a)} \le R\}$ in $Z_1^{(a)c}$, where $\delta = \delta(R) > 0$, such that the functional H^0 is analytic in B^c, $\nabla^a H^0 \in \mathcal{A}_{\mathfrak{A}}^R(B^c; Z_s^{(a)c})$ and

$$\nabla^a H^0(u; a) = \left(\mathcal{A}_a^{-1} \frac{\partial}{\partial w_1} \chi(x, w_1(x); a), 0\right) \tag{6.4}$$

(here ∇^a is the gradient with respect to the scalar product $\langle\cdot;\cdot\rangle^{(a)}$ in the space $Z^{(a)}$).

Proof of analyticity of H^0 and $\nabla^a H^0$ is the same as in Lemma 3.1. The formula for $\nabla^a H^0$ results from the identities

$$\begin{aligned}\langle (v_1, v_2), \nabla^a H^0(w)\rangle^{(a)} &= dH^0(w)(v_1, v_2) = \int \left(\frac{\partial}{\partial w_1}\chi(x, w_1(x); a) v_1(x)\right) dx = \\ &= \int \left(\mathcal{A}_a^{-1} \frac{\partial}{\partial w_1}\chi(x, w_1(x); a)\right) \mathcal{A}_a v_1(x)\, dx \\ &= \langle (v_1, v_2), \left(\mathcal{A}_a^{-1}\frac{\partial}{\partial w_1}\chi(x, w_1(x); a), 0\right)\rangle^{(a)}\ .\end{aligned}$$

□

The Hamiltonian equation corresponding to the hamiltonian $\mathcal{H}_a(w) = \frac{1}{2}\|w\|_0^2 + \varepsilon H^0(w)$ in the symplectic structure given by the 2-form $\langle \bar{J}_a dw, dw\rangle^{(a)}$ has the form

$$(\dot w_1, \dot w_2) = \dot w = J_a \nabla \mathcal{H}_a = \left(\mathcal{A}_a^{1/2} w_2, -\mathcal{A}_a^{1/2}\left(w_1 + \mathcal{A}_a^{-1}\varepsilon \frac{\partial}{\partial w_1}\chi(x, w_1(x); a)\right)\right)\ ;$$

or

$$\dot w_1 = \mathcal{A}_a^{1/2} w_2\ , \quad \dot w_2 = -\mathcal{A}_a^{1/2}\left(w_1 + \mathcal{A}_a^{-1}\varepsilon\frac{\partial}{\partial w_1}\chi(x, w_1(x); a)\right)\ . \tag{6.5}$$

After elimination w_2 from these equations, one gets the second order equation on w_1:

$$\frac{\partial^2}{\partial t^2} w_1 = \left(\frac{\partial^2}{\partial x^2} - V(x; a)\right) w_1 - \varepsilon \frac{\partial}{\partial w_1}\chi(x, w_1(x); a)\ . \tag{6.6}$$

So equation (6.5) is equivalent to (6.1). In what follows we study equation (6.1) in the form (6.5).

As in §3, we have to do some linear transformation before to apply the theorem. So let $\{Z_s\}$ be the scale of Hilbert spaces of the form $\{Z_s^{(a)}\}$ with $V(x; a) \equiv 0$, i.e., defined by the operator $-\partial^2/\partial x^2$ instead of $-\partial^2/\partial x^2 + V(x; a)$. Let us set

$$\varphi_j^+(x) = (\sin jx, 0)(2/\pi j)^{1/2}\ , \quad \varphi_j^- = (0, \sin jx)(2/\pi j)^{1/2}\ ,$$

and denote by $J(a)$ the operator of order 1 in the scale $\{Z_s\}$, such that

$$J(a)\varphi_j^\pm = \mp\left(\lambda_j^{(a)}\right)^{1/2}\varphi_j^\mp\ . \tag{6.7}$$

The triple $\{Z = Z_0, \{Z_s\}, \langle \bar{J}(a)dw, dw\rangle_0\}$ is a symplectic Hilbert scale depending on the parameter a and satisfying relations (1.2).

The linear mapping

$$U_a : Z \longrightarrow Z^{(a)}, \quad \varphi_j^{\pm} \longmapsto \varphi_j^{\pm(a)}$$

transforms one symplectic basis into another and so defines a canonical transformation from Z_s to $Z_s^{(a)}$ for every $s \geq 0$ (see Proposition 2.1′ in Part 1.2). So U_a transforms solutions of the equation

$$\dot{v} = J(a)\big(v + \varepsilon \nabla H(v;a)\big), \quad H(v;a) = H^0(U_a(v);a), \tag{6.8}$$

into solutions of (6.5). It can be checked as in Part 2.3 that each ball in the space Z_1 has a complex neighbourhood $O^c \subset Z_1^c$ such that the functional H is analytic in O^c and $\nabla H \in \mathcal{A}^R_{\mathfrak{A}}(O^c; Z_3^c)$.

Let us fix some compact subset $\mathcal{J}$ of the octant $\mathbb{R}^n_+$ and check that Theorem 1.1 with

$$\lambda_j^A = 1, \quad \lambda_j^J = (\lambda_j^{(a)})^{1/2}, \quad d_J = 1, d_A = 0, d_H = -2, d = 1, d_c = 3,$$

is applicable to equation (6.8). Indeed, assumption 1) means the analyticity of H and ∇H we have just discussed, assumption 2) with $r = 1$ and $K_2 = 1$ holds because $\lambda_j(a) = \lambda_j^J(a) = (\lambda_j^{(a)})^{1/2}$, where $\{\lambda_j^{(a)} = j^2 + C(a) + O(1)\}$ is the spectrum of the Sturm–Liouville problem. Taking into account Remark 2 following Theorem 1.1, we get the following statement about equation (6.5) (or (6.6)).

Theorem 6.2. Let $\rho < 1$ and a_0 be a point in $\mathfrak{A}$ such that

$$D(a_0) = \det\{\partial \lambda_j^{(a_0)}/\partial a_k \mid 1 \leq j,k \leq n\} \neq 0. \tag{6.9}$$

Then there exist integers j_1, M_1 such that if

$$\begin{gathered}\Lambda(a_0;s) = \sqrt{\lambda_1^{(a_0)}}\, s_1 + \sqrt{\lambda_2^{(a_0)}}\, s_2 + \cdots + \sqrt{\lambda_{j_1}^{(a_0)}}\, s_{j_1} \neq 0 \\ \forall s \in \mathbb{Z}^{j_1}, \ |s| \leq M_1, \ 1 \leq |s_{n+1}| + \cdots + |s_{j_1}| \leq 2,\end{gathered} \tag{6.10}$$

then for sufficiently small $\varepsilon > 0$ there exists $\delta_a > 0$, a Borel subset Θ_ε of the set $O(a_0, \delta_a, \mathfrak{A}) \times \mathcal{J}$ and analytic embeddings $\Sigma^\varepsilon_{(a,I)} : \mathrm{T}^n \longrightarrow Z_3$, $(a,I) \in \Theta_\varepsilon$, with the following properties:

a) $\operatorname{mes} \Theta_\varepsilon[I] \longrightarrow \operatorname{mes} O(a_0, \delta_a, \mathfrak{A})$ when $\varepsilon \longrightarrow 0$, uniformly with respect to I;

b) every torus $\Sigma^\varepsilon_{(a,I)}(\mathrm{T}^n) \subset Z_3$ is invariant for the equation (6.5) and is filled with strong in Z_2 quasiperiodic solutions;

c) the tori $\Sigma^\varepsilon_{(a,I)}(\mathrm{T}^n)$, $(a,I) \in \Theta_\varepsilon[I]$, are ε^ρ-close to the invariant tori of the linear equation (6.5) with $\varepsilon = 0$.

The conditions (6.9), (6.10) are the ones of non-degeneracy in the same sense as in §4.

We can as well apply to the equation (6.5) Theorem 2.2 and get the following result:

Theorem 6.3. Let $D(a) \not\equiv 0$ and $\Lambda(a;s) \not\equiv 0$ as a function of a for all $j_1 > n$ and all integer j_1-vectors s such that $1 \leq |s_{n+1}| + \cdots + |s_{j_1}| \leq 2$. Then for every $\gamma > 0$ and for sufficiently small $\varepsilon > 0$ there exists a Borel subset Θ_ε of the set $\Theta_0 = \mathfrak{A} \times \mathcal{J}$ and analytic embeddings

$$\Sigma^{\varepsilon}_{(a,I)} : \mathrm{T}^n \longrightarrow Z_3 , \quad (a,I) \in \Theta_\varepsilon ,$$

such that

a) $\mathrm{mes}(\Theta_0 \backslash \Theta_\varepsilon)[I] < \gamma \ \ \forall I$,

b) every torus $\Sigma^{\varepsilon}_{(a,I)}(\mathrm{T}^n) \subset Z_3$ is invariant for the equation (6.5) and is filled with strong in Z_2 quasiperiodic solutions;

c) the tori $\Sigma^{\varepsilon}_{(a,I)}(\mathrm{T}^n)$ are $C_\gamma\varepsilon$-close to the invariant tori of the unperturbed linear equation.

Example 6.1. Let $n = 1$ and $V(x;a) = a$ for $a \in \mathfrak{A} = (0,1)$. Then $\lambda_m^{(a)} = m^2 + a$ and assumption (6.9) is trivially fulfilled. Let us consider some resonance function of the form (6.10):

$$\Lambda(a;s) = \sum_{m=1}^{j_1} (m^2 + a)^{1/2} s_m . \tag{6.11}$$

The function Λ is analytic in $a \in [0,1]$. It is not identically zero because after the analytic extension into the complex plain with the cut along $(-\infty, 0]$ the function has an essential singularity at the point $a = -m^2$ provided that $s_m \neq 0$. So all the assumptions of Theorem 6.3 are fulfilled and for $j = 1, 2, \ldots$ the time-periodic solutions

$$I \sin jx \sin\left(\sqrt{j^2 + a}\,(t + q)\right) , \quad I > 0 , \quad q \in \mathrm{T}^1 ,$$

of the Klein–Gordon equation

$$w_{tt} = w_{xx} - aw , \quad w(t,0) \equiv w(t,\pi) \equiv 0 ,$$

persist in the nonlinear Klein–Gordon equation

$$w_{tt} = w_{xx} - aw - \varepsilon \frac{\partial}{\partial w} \chi(x,w) , \quad w(t,0) \equiv w(t,\pi) \equiv 0$$

for most of $a \in (0,1)$ and ε small enough. As above, the function χ is smooth in x, analytic in w and $\chi(0,0) = \chi(\pi,0) = 0$.

Example 6.2 Let $n \geq 2$, $a = (a_1, \ldots, a_n) \in \mathfrak{A}$, where

$$\mathfrak{A} = \left\{ a \in O(1, \mathrm{R}^n) \;\middle|\; a_1 \pm a_2 \pm \cdots \pm a_n > 0 \quad \text{for all choices of } \pm \right\} ,$$

and

$$V(x;a) = a_1 - a_2 \cos 4x - \cdots - a_n \cos 2nx \tag{6.12}$$

(thus $V(x;a) > 0$ for all x, a). Let us numerate the eigen-functions $\varphi_j(x;a)$ of the operator $-\partial/\partial x^2 + V(x;a)$ in such a way that $\varphi_j(x;0) = (2/\pi)^{1/2} \sin(jx)$ for all j.

As $V(x;a_1,0,\dots,0) = a_1$, then

$$\Lambda(a_1,0,\dots,0;s) = \Lambda(a_1;s)$$

with the function $\Lambda(a;s)$ as in (6.11). So for nonzero s $\Lambda(a_1,0,\dots,0;s) \not\equiv 0$, as it was shown in Example 6.1. Thus, $\Lambda(a,s) \not\equiv 0$.

To check the assumption $D(a) \not\equiv 0$, let us calculate $D(a)$ at the limiting point $a = 0$. By formula (3.15),

$$\frac{\partial \lambda_j}{\partial a_1}(0) = \int_0^\pi \varphi_j^2(x;a_0)\frac{\partial V(x;a_0)}{\partial a_1}\,dx = \frac{2}{\pi}\sin^2 jx\,dx = 1\,;$$

and for $k \geq 2$

$$\frac{\partial \lambda_j}{\partial a_k}(0) = \frac{2}{\pi}\int_0^\pi \sin^2 jx \cos(2kx)\,dx = \frac{1}{2}\delta_{j,k}\,.$$

Thus, $D(0) = 2^{1-n}$. So $D(x) \not\equiv 0$ and the assumptions of Theorem 6.3 hold for the potential $V(x;a)$ of the form (6.12).

7. On non-commuting operators J, A and partially hyperbolic invariant tori

The assumptions (1.2), (1.3) that the operators A^Z and J^Z in (1.6) commute and are diagonal in some Hilbert basis hold for many applications (e.g., for the ones given in §§3–6 above), but look rather restrictive. In fact, these assumptions are not a restriction at all, because in the elliptic situation they can be easily achieved by means of an additional symplectic linear transformation of the phase space Z, provided that the eigenfunctions of the operator $J^Z A^Z$ "asymptotically behave in a regular way". In the partially hyperbolic situation the same linear transformation put the linear equation (1.6) to a normal form, similar to (1.2), (1.3). Below we present a scheme of this reduction and give a version of Theorem 1.1, which is applicable in the partially hyperbolic situation.

We suppose that the antiselfadjoint in Z operator $J^Z(a)$ and selfadjoint operator $A^Z(a)$ define isomorphisms of the scale $\{Z_s\}$ of orders $d_J \geq 0$, $d_A \geq 0$, and that these operators satisfy (1.0). We also suppose that the operator $J^Z(a)$ has the form (1.2) in some Hilbert basis $\{\varphi_j^\pm\}$ ((1.2) is the canonical form for an antiselfadjoint operator in a finite-dimensional Hilbert space, in the infinite-dimensional setting it means some regularity of this operator).

We denote $D = D(a) = J^Z(a)A^Z(a)$.

Lemma 7.1. Spectrum $\sigma = \sigma(a)$ of the operator D is symmetric with respect to the involutions $\lambda \longmapsto -\lambda$, $\lambda \longmapsto \bar\lambda$.

Proof. If $\lambda \notin \sigma$, then the operator $D - \lambda$ is invertible in Z^c. As $(D - \bar\lambda)^{-1} = \overline{(D-\lambda)^{-1}}$, then also $\bar\lambda \notin \sigma$. So $\bar\sigma = \sigma$.

To prove that $-\sigma = \sigma$ we consider any $\lambda \notin \sigma$. As we have seen, $\bar{\lambda} \notin \sigma$. Hence the point λ also lies outside the spectrum of the adjoint operator $D^* = (J^Z A^Z)^* = -A^Z J^Z$. So the operator $-A^Z J^Z - \lambda$ is invertible, as well as the operator $-J(-AJ - \lambda)J^{-1} = JA + \lambda$. Thus $-\lambda \notin \sigma$ and the lemma is proven. □

As it is always the case in this book, we suppose that the operator D has pure point spectrum. By Lemma 7.1 the spectrum can be written as $\sigma = \{\pm i\lambda_j | j \in \mathbb{N}\}$. We also suppose that the numbers $\lambda_j = \lambda_j(a)$ are real positive for all $j \geq j_*$ with some j_* large enough, and that the spectrum σ is single (the latter holds under the assumptions of Theorem 1.1 if the radius δ_a is small enough). By (1.0) the operator D is invertible, so $0 \notin \sigma$.

So almost all points in σ are pure imaginary. We remind that the zero equilibrium point of the equation (1.6) is called *elliptic*, if $\sigma \subset i\mathbb{R}$, and *partially hyperbolic*, if some finite system of eigenvalues $\pm i\lambda_j$ lie outside of the imaginary axis $i\mathbb{R}$. For short we shall say that the equation (1.6) itself is elliptic or partially hyperbolic.

Consider complex eigen-functions $\{W_j^\pm(a)\}$ of the operator $D(a)$:

$$DW_j^\pm = \pm i\lambda_j W_j^\pm \,, \quad \pm i\lambda_j \in \sigma \,.$$

The function $\overline{W_j^+}$ is an eigenfunction with the eigenvalue $-i\bar{\lambda}_j$. As the spectrum σ is single, the vector $\overline{W_j^+}$ is proportional to the eigenvector W_k^-, where $\bar{\lambda}_j = \lambda_k$. So the eigenfunctions can be chosen in such a way that

$$\overline{W_j^+} = W_k^- \quad \text{if} \quad \bar{\lambda}_j = \lambda_k \,. \tag{7.1}$$

We suppose that the vectors $\{W_j^\pm\}$ form a basis of the complex space Z^c and that they are asymptotically close to the vectors of the standard complex basis

$$\{w_j^\pm = (\varphi_j^+ \pm i\varphi_j^-)/\sqrt{2}\}$$

(we do not state the last requirement exactly and just note that it is fulfilled if D is a lower-order perturbation of some constant-coefficient differential operator $J_0 A_0$, where J_0 and A_0 meet assumptions (1.2), (1.3)).[11]

As in the finite-dimensional situation, eigenvectors are skew-orthogonal unless the sum of their eigenvalues vanish:

$$\alpha[W_j^+, W_k^\pm] = 0 \quad \text{if} \quad \lambda_j \pm \lambda_k \neq 0 \,. \tag{7.2}$$

Indeed,

$$\alpha[W_j^+, W_k^\pm] = \langle \bar{J}^Z W_j^+, W_k^\pm \rangle = \frac{1}{i\lambda_j} \langle \bar{J}^Z J^Z A^Z W_j^+, W_k^\pm \rangle = -\frac{1}{i\lambda_j} \langle W_j^+, A^Z W_k^\pm \rangle$$
$$= \frac{1}{i\lambda_j} \langle W_j^+, \bar{J}^Z J^Z A^Z W_k^\pm \rangle = -\frac{\pm i\lambda_k}{i\lambda_j} \alpha[W_j^+, W_k^\pm] \,.$$

---

[11] In Lemma 3.2 above the properties we need were checked for the operator $J^Z A^Z = i\big(-\partial^2/\partial x^2 + V(x)\big)$.

This equality clearly implies (7.2).

For $j \in \mathbb{N}$ we define $\mathcal{Z}^c(j)$ as the complex linear span in Z^c of the eigenvectors with the eigenvalues $\pm i\lambda_j$ and $\pm i\bar{\lambda}_j$ (the dimension of this space equals 2 if $\lambda_j \in \mathbb{R} \cup i\mathbb{R}$ and equals 4 otherwise). By (7.1) this space is the complexification of the real space $\mathcal{Z}(j) = \mathcal{Z}^c(j) \cap Z$. Different spaces $\mathcal{Z}^c(j)$ are indexed by $j \in \mathcal{N} = \mathcal{N}_1 \cup \mathcal{N}_2 \cup \mathcal{N}_3 \subset \mathbb{N}$, where the sets $\mathcal{N}_j$ are defined by the following relations:

$$j \in \begin{cases} \mathcal{N}_1 , & \text{if } \lambda_j \in \mathbb{R} , \\ \mathcal{N}_2 , & \text{if } \lambda_j \in i\mathbb{R} , \\ \mathcal{N}_3 , & \text{if } \lambda_j \notin \mathbb{R} \cup i\mathbb{R} . \end{cases}$$

By our assumption, $j \in \mathcal{N}_1$ if $j \geq j_*$. So the sets of hyperbolic directions $\mathcal{N}_2$ and $\mathcal{N}_3$ are finite. The sets $\mathcal{N}_1, \mathcal{N}_2, \mathcal{N}_3$ are a-independent because the spectrum $\sigma = \sigma(a)$ is single for all a and the set $\mathfrak{A}$ is connected.

By (7.2) different spaces $\mathcal{Z}^c(j)$ are skew-orthogonal. As the vectors $\{W_j^\pm\}$ form a basis of Z^c, then Z^c is the skew orthogonal direct sum of these subspaces:

$$Z^c = \bigoplus_{j \in \mathcal{N}} \mathcal{Z}^c(j) .$$

In particular, $\mathcal{Z}^c(j)$ are symplectic subspaces of Z^c.

For elliptic directions $j \in \mathcal{N}_1$ we have from (7.1) $\overline{W_j^+} = W_j^-$. So the numbers $\alpha[W_j^+, W_j^-]$ are imaginary nonzero. They can be normalized in such a way that

$$\alpha[W_j^+, W_j^-] = \alpha[w_j^+, w_j^-] \quad \forall j \in \mathcal{N}_1 .$$

Define the spaces

$$Z^{c1} = \overline{\text{span}}\{W_j^\pm \mid j \in \mathcal{N}_1\} , \quad Z^{c2} = \text{span}\{W_j^\pm \mid j \in \mathcal{N}_2 \cup \mathcal{N}_3\} ,$$

and define the spaces $\mathcal{Z}^{c1}, \mathcal{Z}^{c2}$ in the same way but with the vectors $W_j^\pm$ replaced by $w_j^\pm$. For $k = 1, 2$ denote by J_k the restriction of the operator $J^Z(a)$ to Z^{ck}. The linear map

$$L_1 : Z^{c1} \longrightarrow \mathcal{Z}^{c1} , \quad w_j^\pm \longmapsto W_j^\pm \quad \forall j \in \mathcal{N}_1$$

is symplectic and real (it maps real vectors to real). It transforms equation (1.6), restricted to $\mathcal{Z}^{c1}$, to the equation

$$\dot{u} = J_1 \hat{A}_1 u , \quad u \in Z^{c1} ,$$

where for $j \in \mathcal{N}_1$

$$\hat{A}_1 \varphi_j^\pm = \lambda_j^A \varphi_j^\pm , \quad \lambda_j^A = \lambda_j / \lambda_j^J .$$

The symplectic spaces Z^{c2} and $\mathcal{Z}^{c2}$ have the same dimension. So they are symplectomorphic and $L_2 : Z^{c2} \longrightarrow \mathcal{Z}^{c2}$ for some real symplectomorphism L_2. The transformed linear equation in $\widetilde{Z}^{c2}$ has the form

$$\dot{u} = J_2 \hat{A}_2 u , \quad u \in Z^{c2} ,$$

with some real selfadjoint operator $\hat{A}_2$. The operator $J_2\hat{A}_2$ has the hyperbolic spectrum $\{\pm i\lambda_j(a) | j \in \mathcal{N}_2 \cup \mathcal{N}_3\}$ and the transformation L_2 can be chosen in such a way that the eigenvectors of $J_2\hat{A}_2$ are a-independent.

Define

$$L = L_1 \oplus L_2 : Z^c = Z^{c1} \oplus Z^{c2} \longrightarrow Z^c = Z^{c1} \oplus Z^{c2} .$$

This map transforms (1.6) to the linear equation

$$\dot{u} = J^Z(a)\hat{A}^Z(a)u \ , \quad u \in Z \ ,$$

where $\hat{A}^Z = \hat{A}_1 \oplus \hat{A}_2$.

In particular, if the equation (1.6) is elliptic (i.e., all the eigenvalues $\pm i\lambda_j$ are imaginary), then $Z^{c2} = \{0\}$ and (1.6) is reduced to an equation of the same form, but with the operator $A^Z = \hat{A}^Z$ satisfying (1.3).

If the set $\mathcal{N}_2 \cup \mathcal{N}_3$ of the hyperbolic directions is not empty, then we are in more general situation, than the one considered in Theorem 1.1. Still, if $\{1, 2, \ldots, n\} \subset \mathcal{N}_1$, then the tori $T(I)$ as in §1 are well-defined and are invariant for the linear equation (1.6). The same proof we give below for Theorem 1.1, can be also applied to this situation (but with more cumbersome notations, which is the reason why we present a proof for the elliptic situation only). Thus obtaining result can be stated as follows:

Theorem 7.2. Let all the assumptions of Theorem 1.1 hold, except the assumption (1.3) which holds only for $j \leq n$ and for $j > j_*$ with some $j_* \geq n$. Let for $j = n+1, \ldots, j_*$ the operator $J^Z(a)A^Z(a)$ has a-independent complex eigenvectors $W_j^{\pm}$, corresponding to the eigenvalues $\pm i\lambda_j(a) \notin i\mathbb{R}$, which together span the space $\mathbb{C}\varphi_{n+1}^+ + \mathbb{C}\varphi_{n+1}^- + \cdots + \mathbb{C}\varphi_{j_*}^+ + \mathbb{C}\varphi_{j_*}^-$. Then statements a) – c) of Theorem 1.1 remain true. The quasiperiodic solutions $z^\varepsilon(t)$ are partially hyperbolic and the number of hyperbolic directions is equal to $2(|\mathcal{N}_2| + |\mathcal{N}_3|) = 2(j_* - n)$.

Example. Let us consider the problem (6.6), (6.2) without the restriction $V(x;a) \geq 0$ (and, so, with the possibility of $p \geq 0$ negative points in the spectrum $\sigma(\mathcal{A}_a)$ of the operator $\mathcal{A}_a = -\partial^2/\partial x^2 + V(x;a)$). Let us suppose that the operator $\mathcal{A}_a$ is invertible and denote by $\mathcal{Z}_t$, $t \geq 0$, the space $\mathcal{Z}_t = D(|\mathcal{A}_a|^{(t+1)/2})$. Let us define the spaces $\{Z_s^{(a)}\}$, $\{Z_s\}$, the operators J_a, $J(a)$ and the function H^0 in the same way as in §6, but with the operator $|\mathcal{A}_a|$ instead of $\mathcal{A}_a$ and $\left|\lambda_j^{(a)}\right|$ instead of $\lambda_j^{(a)}$, $j = 1, 2, \ldots$ (by definition, $|\mathcal{A}_a| \varphi_j(a) = \left|\lambda_j^{(a)}\right| \varphi_j(a) \ \ \forall j$). Let us consider the hamiltonian

$$\mathcal{H}(w;a) = \frac{1}{2}\|w\|^2 + \frac{1}{2}\langle(\operatorname{sgn}\mathcal{A}_a)w^1, w^1\rangle + \varepsilon H^0(w;a) \ , \tag{7.3}$$

where

$$\operatorname{sgn}\mathcal{A}_a \varphi_j(a) = \operatorname{sgn}\lambda_j^{(a)} \varphi_j(a) \ \ \forall j \ .$$

The corresponding Hamiltonian equations have the form

$$\begin{aligned} \dot{w}^1 &= |\mathcal{A}_a|^{1/2}\, w^2\,, \\ \dot{w}^2 &= -\,|\mathcal{A}_a|^{1/2}\Big((\operatorname{sgn}\mathcal{A}_a)w^1 + \varepsilon\,|\mathcal{A}_a|^{-1}\,\frac{\partial}{\partial w^1}\chi\Big)\,, \end{aligned} \tag{7.4}$$

and we get the equation (6.6) for the function $w^1(t,x)$, again.

Let us numerate the eigenfunctions $\{\varphi_j(x;a)\}$ and the eigenvalues $\{\lambda_j^{(a)}\}$ of the operator $\mathcal{A}_a$ in such a way that

$$\lambda_{n+1}^{(a)},\dots,\lambda_{n+p}^{(a)} < 0 < \lambda_1^{(a)},\dots,\lambda_n^{(a)},\lambda_{n+p+1}^{(a)},\dots\,.$$

Then the linear equation $(7.4)|_{\varepsilon=0}$ has $2p$ hyperbolic directions with the eigenvalues $\pm\sqrt{\lambda_{n+1}^{(a)}},\dots,\pm\sqrt{\lambda_{n+p}^{(a)}}$. So $|\mathcal{N}_2| = p$ and $|\mathcal{N}_3| = 0$. By the same arguments as we applied in Part 6 to the equation (6.5), Theorem 7.2 with $j_* = n+p-1$ is applicable to study equation (7.4) with a "typical" potential $V(x;a)$. So the assertions of Theorems 6.2, 6.3, concerning the nonlinear string equation, remain true without the assumption $V(x;a) \geq 0$.

Thus the nonlinear string equation with a negative potential gives a natural example of equation (1.5) such that the unperturbed linear equation (1.6) is partially hyperbolic with real hyperbolic eigenvalues. Nonlinear perturbations of partially hyperbolic linear equations (1.6) with the hyperbolic eigenvalues $\pm i\lambda_j \notin \mathbb{R} \cup i\mathbb{R}$ arise in study the Sine–Gordon equation in the vicinity of its large-amplitude finite-gap solutions. We do not discuss here this rather extended subject and refer the reader to [BiK] (see also [EFM]).

Appendix. On superposition operator in Sobolev spaces

Let $O^c \subset \mathbb{C}$ be a complex neighborhood of the real line and $\chi : O^c \times [0,\pi] \to \mathbb{C}^p$ be a C^k-function which is real for real arguments. Let B_R be the ball in the Sobolev space $\mathfrak{Z}_k = H^k(0,\pi;\mathbb{R}^p)$ of radius R, centered at zero, and $B_R^c(\delta)$ be its δ-neighbourhood in the complex space $\mathfrak{Z}_k^c = H^k(0,\pi;\mathbb{C}^p)$. If $k \geq 1$, then $B_R^c(\delta) \subset C(0,\pi;O^c)$ provided that $\delta = \delta(R)$ is small enough. So the superposition operator

$$\varphi : B_R^c(\delta) \longrightarrow C(0,\pi;\mathbb{C}^p)\,, \quad u(x) \longmapsto \chi(u(x),x)\,,$$

is well-defined.

Theorem A1. Let us suppose that C^k-smooth function $\chi(u,x)$ is analytic in u:

$$\frac{\partial^s}{\partial x^s}\chi(\cdot,x) \in \mathcal{A}^R(O_B^c;\mathbb{C}^p) \quad \text{and}$$

$$\sup_{u\in O_B^c}\left|\frac{\partial^s}{\partial x^s}\chi(u,x)\right| \leq K = K(O_B^c) \quad \forall x\,, \quad \forall s \leq k\,,$$

for each bounded subdomain $O_B^c \Subset O^c$. Then the superposition operator φ is analytic,

$$\varphi \in \mathcal{A}^R(B_R^c(\delta);\mathfrak{Z}_k^c)$$

and

$$|\varphi(u)|_{\mathfrak{Z}_k^c} \leq KC(R) \quad \forall u \in B_R^c(\delta)\,. \tag{A1}$$

Proof. By taking derivatives of order $\ell \leq k$ from the function $\chi(u(x),x)$ with $u \in B_R^c(\delta)$, one gets the estimate (A1). If $u \in B_R^c(\delta)$ and $v,w \in \mathfrak{Z}_k^c$, then the function

$$\lambda \longmapsto \langle \varphi(u+\lambda v), w\rangle_{\mathfrak{Z}_k^c}$$

is complex-analytic in some neighborhood of the origin in $\mathbb{C}$; so the map φ is weakly analytic in $B_R^c(\delta)$. As φ is bounded and weakly analytic, then it is Fréchet-analytic (see e.g., [PT], Appendix A). □

Let the function $\chi = \chi(u,x;a)$ depends on a parameter $a \in \mathfrak{A}$ in the Lipschitzian way, i.e., $\chi(\cdot,\cdot;a) \in C^k(O^c \times [0,\pi])$ for all a, and

$$\frac{\partial^s}{\partial x^s}\chi(\cdot,x;\cdot) \in \mathcal{A}^R_{\mathfrak{A}}(O_B^c;\mathbb{C}^p) \quad \forall s \leq k, \quad \forall x \in [0,\pi]\,, \tag{A2}$$

for each $O_B^c \Subset O^c$. Then by applying Theorem A1 to the functions $\chi(u(x),x,a)$ and $\big(\chi(u(x),x,a_1) - \chi(u(x),x,a_2)\big)$ we get

Corollary A2. If assumption (A2) holds with some $k \geq 1$, then the map φ belongs to $\mathcal{A}^R_{\mathfrak{A}}(B_R^c(\delta);\mathfrak{Z}_k^c)$. In particular, the function $u(x) \longmapsto \int \varphi(u)(x)\,dx$ belongs to $\mathcal{A}^R_{\mathfrak{A}}(B_R^c(\delta);\mathbb{C}^p)$.

Part 3

Proof of the main theorem

We use the notations from Parts 1, 2 and some new ones. A list of the notations is given at the end of the paper. Sometimes we refer the reader to the formulas from Part 2. We write (2.2.3) for the formula (2.3) from Part 2 and so on. We use the abbreviations r.h.s. (l.h.s.) for "right-hand-side"("left-hand-side") and write ε_0 instead of ε. By $\varepsilon_0 << 1$ ($K >> 1$) we mean positive ε_0 is small enough (K is large enough).

1. Preliminary transformations

In a symplectic Hilbert scale $\{Z, \{Z_s | s \in \mathbb{R}\}, \alpha(a) = \langle \overline{J^Z}(a)\, dz, dz\rangle_Z\}$ (see Part 1) we study the Hamiltonian equation with the hamiltonian

$$\mathcal{H}(z; a, \varepsilon_0) = \frac{1}{2}\langle A^Z(a)z, z\rangle_Z + \varepsilon_0 H(z; a, \varepsilon_0)\,;$$

i.e. the equation

$$\dot{z} = J^Z(a)\big(A^Z(a)z + \varepsilon_0 \nabla H(z; a, \varepsilon_0)\big)\,, \quad J^Z(a) = -\big(\overline{J^Z}(a)\big)^{-1}\,. \tag{1.1}$$

Here $a \in \mathfrak{A} \Subset \mathbb{R}^n$ is a n-dimensional parameter, $\varepsilon_0 \in [0,1]$ is a small parameter, H is an analytical function, $J^Z(a)$, $A^Z(a)$ are linear operators, defining isomorphisms of the scale $\{Z_s\}$ of orders d_J, d_A and such that for some Hilbert basis $\{\varphi_j^\pm | j \geq 1\}$ of the space Z the following relations hold for all j and a:

$$J^Z(a)\varphi_j^\pm = \mp\lambda_j^J(a)\varphi_j^\mp\,, \quad A^Z(a)\varphi_j^\pm = \lambda_j^A(a)\varphi_j^\pm\,.$$

For the exact assumptions on equation (1.1) see Part 2.

1.1. Change of symplectic structure

All the numbers $\{\lambda_j^J(a)\}$ are nonzero and are positive if j is large enough. So after unessential exchange $\varphi_j^\pm$ on $\varphi_j^\mp$ for some finite number of indexes j we may suppose that $\lambda_j^J(a)$ are positive for all j. Let us consider the linear operator L_a which maps $\varphi_j^\pm$ to $\big(\lambda_j^J(a)\big)^{1/2}\varphi_j^\pm$, $j = 1, 2, \ldots$. This operator defines an isomorphism of the scale $\{Z_s\}$ of order $d_J/2$. Clearly, the operator is selfadjoint in Z with the domain of definition $Z_{d_J/2}$. So by Corollary 2.3 from Part 1 the mapping L_a^{-1} transforms solutions of the Hamiltonian equation (1.1) in the symplectic

Hilbert scale $\{Z, \{Z_s\}, \alpha(a)\}$ into solutions of the Hamiltonian equation with the hamiltonian

$$\mathcal{H}_1(z; a, \varepsilon_0) = \frac{1}{2}\langle A_1(a)z, z\rangle_Z + \varepsilon_0 H_1(z; a, \varepsilon_0)$$

in the symplectic Hilbert scale $\{Z, \{Z_s\}, \alpha_1(a) = \langle \bar{J}_1(a)\, dz, dz\rangle_Z\}$. Here

$$\bar{J}_1(a) = L_a \bar{J}^Z(a) L_a \,, \quad A_1(a) = L_a A^Z(a) L_a \,, \quad H_1 = H(L_a z; a, \varepsilon_0) \,.$$

By the definition of the operator L_a we see that for all j and a we have $\bar{J}_1(a)\varphi_j^{\pm} = \mp\varphi_j^{\mp}$. So the operator $\bar{J}_1(a)$ does not depend on the parameter a and $J_1 = -(\bar{J}_1)^{-1} = \bar{J}_1$.

We denote by $\mathcal{L}(Z_s; Z_{s_1})$ the space of linear continuous operators from Z_s to Z_{s_1} endowed with the operator norm $\|\cdot\|_{s,s_1}$, and consider the map $L : \mathfrak{A} \to \mathcal{L}(Z_s; Z_{s-d_J/2})$ which corresponds the operator L_a to a point $a \in \mathfrak{A}$.

Lemma 1.1. For every s

$$\operatorname{Lip}\big(L : \mathfrak{A} \to \mathcal{L}(Z_s; Z_{s-d_J/2})\big) \le C \,.$$

For every s and every a

$$\|L_a\|_{s,s-d_J/2} + \left\|L_a^{-1}\right\|_{s,s+d_J/2} \le C_1 \,.$$

Proof. The operator $L_{a_1} - L_{a_2}$ is diagonal in the basis $\{\varphi_j^{\pm}\}$ with the eigenvalues $\Delta\ell_j^{\pm} = \left(\lambda_j^J(a_1)\right)^{1/2} - \left(\lambda_j^J(a_2)\right)^{1/2}$. By the assumptions (2.1.0) – (2.1.2) and (2.1.12),

$$|\Delta\ell_j^{\pm}| \le \frac{K_1\,|a_1 - a_2|\,j^{d_J}}{2\min(\lambda_j^J(a_1)^{1/2}, \lambda_j^J(a_2)^{1/2})} \le \frac{K_1^{3/2}}{2}\,|a_1 - a_2|\,j^{d_J/2} \,,$$

and we get the first estimate. The second one results from (2.1.0) – (2.1.2). □

Let us denote $d' = d + \frac{1}{2}d_J$. The map L_a^{-1} transforms the tori $T(I)$ to n-tori $T_a(I)$ in $Z_{d'}$, where a torus $T_a(I)$, $I \in \mathcal{J}$, has the same form as $T(I)$, but with I_j replaced by $I_j^a = I_j/\lambda_j^J(a)$ for all j. We denote by $\mathcal{T}_a$ the union of these tori:

$$\mathcal{T}_a = \bigcup\{T_a(I) \mid I \in \mathcal{J}\} \,.$$

By the assumption 1) of the theorem and by Lemma 1.1, the function H_1 may be analytically extended to some δ-neighborhood of $\mathcal{T}_a$ in $Z_{d'}^c$. In this neighborhood $|H_1|$ and $\|\nabla H_1\|_{d'-d_H-d_J}$ are bounded together with the corresponding Lipschitz constants.

The operator $A_1(a)$ is diagonal in the basis $\{\varphi_j^{\pm}\}$ and $A_1(a)\varphi_j^{\pm} = \lambda_j(a)\varphi_j^{\pm}$ for all j and a. So the transformed equation meets the theorem's assumptions with $d'_A = d_A + d_J$ and $d'_J = 0$.

We have seen that it is sufficient to prove the theorem in the case $d_J = 0$.

1.2. Change of parameter

The statements of the theorem are local with respect to the parameter a. So we can replace the set $\mathfrak{A}$ of parameters a by arbitrary δ_a-neighborhood $\mathfrak{A}(a_0, \delta_a)$ of the point a_0 in $\mathfrak{A}$. If positive δ_a is sufficiently small, then by the assumptions (2.1.7), (2.1.8) the mapping

$$\omega : \mathfrak{A}(a_0, \delta_a) \longrightarrow \mathbb{R}^n , \quad a \longmapsto \omega(a) = (\lambda_1(a), \ldots, \lambda_n(a))$$

defines a C^1-diffeomorphism on some neighborhood Ω_0 of the point $\omega_0 = \omega_0(a_0)$ and

$$\operatorname{Lip} \omega + \operatorname{Lip} \omega^{-1} \leq K^1 .$$

So the Lipschitz dependence on the parameter $a \in \mathfrak{A}(a_0, \delta_a)$ is equivalent to the Lipschitz dependence on the parameter $\omega \in \Omega_0$. We somewhat generalize the situation and suppose below that $\Omega_0 \subset \mathbb{R}^n$ is a Borel set such that

$$\operatorname{diam} \Omega_0 \leq K\delta_a , \quad \operatorname{mes} \Omega_0 \geq K^{-1}\delta_a^n , \quad |\omega| \leq K \quad \forall \omega \in \Omega_0 \tag{1.2}$$

with some $0 < \delta_a \leq 1$.

1.3. Transition to angle-action variables

We remind that by $O(Q, \delta, B)$ we denote the δ-neighborhood of a subset Q of a metric space B; for a Banach space Z we write $O(\delta, Z)$ instead of $O(0, \delta, Z)$.

Let us set $Z^0 \subset Z$ be equal to the $2n$-dimensional linear span of the vectors $\{\varphi_j^\pm | j \leq n\}$ and $Y_s \subset Z_s$, $s \in \mathbb{R}$, be equal to the closure in Z_s of a linear span of the vectors $\{\varphi_j^\pm | j \geq n+1\}$. We denote $Y = Y_0$ and provide Y with the scalar product $\langle \cdot, \cdot \rangle = \langle \cdot, \cdot \rangle_Y$, induced from Z.

For a vector from Z^0 let $\{\chi_j^\pm | 1 \leq j \leq n\}$ be its coefficients with respect to the basis $\{\varphi_j^\pm | j \leq n\}$. In some small enough neighborhood of a torus $T_a(I)$ let us change coordinates $\{\chi_j^\pm\}$ to the angle-action variables (q, ξ), where $q \in T^n$ and $\xi \in \mathbb{R}^n$, $|\xi| < \delta$, with $\delta << 1$:

$$q_j = \operatorname{Arg}(\chi_j^- + i\chi_j^+) , \qquad \xi_j = \frac{1}{2}\left[\chi_j^{+^2} + \chi_j^{-^2}\right] - I_j^a .$$

The variables (q, ξ, y) vary in the toroidal space $\mathcal{Y}_s = T^n \times \mathbb{R}^n \times Y_s$ endowed with the natural metric dist_s. For all s the map

$$(q, \xi, y) \longmapsto \sum \chi_j^\pm \varphi_j^\pm + y \tag{1.3}$$

defines a complex-analytic diffeomorphism of the δ_0-neighborhood $Q^c(s)$ of the torus $T_0^n = T^n \times \{0\} \times \{0\}$ in the complexification $\mathcal{J}_s^c$ of the phase-space $\mathcal{J}_s$,

$$\mathcal{J}_s^c = \{\mathbb{C}^n / 2\pi \mathbb{Z}^n\} \times \mathbb{C}^n \times Y_s^c ,$$

and a neighborhood of $T_a(I)$ in Z^c_s. This diffeomorphism is Lipschitz in I and ω.

The tangent spaces $T_u\mathcal{Y}_s$, $u \in \mathcal{Y}_s$, are isomorphic to the Hilbert space $E^s = \mathbb{R}^n \times \mathbb{R}^n \times Y_s$. Let J be a restriction on Y_s of the operator J_1, let J^T be the map of $\mathbb{R}^n \times \mathbb{R}^n$ into itself, such that $J^T(q,\xi) = (\xi, -q)$, and

$$J^{\mathcal{Y}} = J^T \times J^Y : E_s = (\mathbb{R}^n \times \mathbb{R}^n) \times Y_s \longrightarrow E_s \,.$$

We introduce in $\mathcal{Y}_s$, $s \geq 0$, a symplectic structure with the help of the 2-form $\alpha^{\mathcal{Y}} = \langle J^{\mathcal{Y}} d\eta, d\eta\rangle_E$. The triple $\{\mathcal{Y}_0, \{\mathcal{Y}_s\}, \alpha^{\mathcal{Y}}\}$ is a toroidal symplectic Hilbert scale (see Part 1, §4) and the map (1.3) defines a symplectomorphism with respect to this symplectic structure. In the coordinates (q,ξ,y) the equation (1.1) is Hamiltonian with the transformed hamiltonian

$$\mathcal{H}_0(q,\xi,y;\omega,I,\varepsilon_0) = \text{const} + \sum_{j=1}^{n} \xi_j\omega_j + \frac{1}{2}\langle A(\omega)y,y\rangle + \varepsilon_0 H^0(q,\xi,y;\omega,I,\varepsilon_0)$$

and has the form

$$\dot q = \nabla_\xi \mathcal{H}_0 \,, \quad \dot\xi = -\nabla_q \mathcal{H}_0 \,, \quad \dot y = J^Y \nabla_y \mathcal{H}_0 \tag{1.4}$$

(see Part 1, Proposition 4.1). Here we denote by $A(\omega)$ the restriction of the operator $A_1(\omega)$ to the space Y and use the fact that the restriction of the quadratic form $\frac{1}{2}\langle A_1(\omega)z,z\rangle_Z$ to Z^0 is equal to $\omega_1\xi_1 + \cdots + \omega_n\xi_n$.

1.4. Reformulation of the theorem

The hamiltonian $\mathcal{H}_0$ depends on parameter $(\omega,I) \in \Theta_0 = \Omega_0 \times \mathcal{J}$, where the domain Ω_0 is as in (1.2). We carry out one more generalization and suppose below that $\mathcal{J}$ is an arbitrary metric space (possibly, $\mathcal{J}$ contains the only point), endowed with the σ-algebra of Borel sets. We denote pairs (ω,I) by θ. So

$$\Theta_0 = \Omega_0 \times \mathcal{J} = \{\theta = (\omega,I)\} \,.$$

The hamiltonian H^0 may be analytically extended to some δ-neighborhood $Q^c(d')$ or the torus $T_0^n = \mathbb{T}^n \times \{0\} \times \{0\}$ in $\mathcal{Y}^c_{d'}$ and is there Lipschitz in (ω, I):

$$\left|H^0(\cdot;\cdot,\varepsilon_0)\right|^{Q^c(d'),\Theta_0} \leq K_1', \quad \left\|\nabla_y H^0(\cdot;\cdot,\varepsilon_0)\right\|^{Q^c(d'),\Theta_0}_{d-d_H-d_J} \leq K_1' \,.$$

The operator $A(\omega)$ has the double spectrum $\{\lambda_j(\omega) | j = n+1, n+2, \dots\}$ and the operator $JA(\omega)$ has the spectrum $\{\pm i\lambda_j(\omega) | j \geq n+1\}$. Let us shift the numeration:

$$\lambda_j(\omega) := \lambda_{j+n}(\omega) \,, \quad \varphi_j^{\pm} := \varphi_{j+n}^{\pm} \,, \quad \lambda_j^{(s)} := \lambda_{j+n}^{(s)} \,,$$

and redenote

$$d_H := d_H + d_J \,, \quad d := d' = d + \frac{1}{2} d_J \,.$$

Then by the condition (2.1.2) for any real s the set of vectors $\{\varphi_j^{\pm}\lambda_j^{(-s)}|j\geq 1\}$ forms a Hilbert basis of the space Y_s and

$$K^{-1}j^s \leq \lambda_j^{(s)} \leq Kj^s, \ \ \lambda_j^{(s)} = (\lambda_j^{(s)})^{-1} \ \ \forall j \geq 1\,, \tag{1.5}$$

with some new constant K. By these properties the Hilbert scale $\{Y_s\}$ is interpolational (see below appendix A).

For the shifted sequence $\{\lambda_j\}$ the relation (2.1.11) holds with the same d_1, some new numbers r, K_2^ℓ, $d_{1,\ell}$ and new K^1. Below the frequencies λ_j (but not the coefficients K_2^ℓ) are allowed to depend on $\theta=(\omega,I)\in\Theta_0$ and ε_0, so for $j\geq 1$

$$A(\theta,\varepsilon_0)\varphi_j^{\pm} = \lambda_j(\theta,\varepsilon_0)\varphi_j^{\pm}, \ \ J\varphi_j^{\pm} = \mp\varphi_j^{\mp}\,. \tag{1.6}$$

Now we reformulate Theorem 1.1 from Part 2 for equations (1.4). In fact, we state (and prove) some more general result, which suits better to the mentioned in the introduction applications to nonlinear equations. So we suppose that the perturbation $\varepsilon_0 H^0$ is of more general form than above and is equal to

$$\varepsilon_0 H^0 = \varepsilon_0 H_0(q,\xi,y;\theta,\varepsilon_0) + H^3(q,\xi,y;\theta,\varepsilon_0)\,,$$

where the functions H_0, H^3 depend on ε_0 and may be analytically (in (q,ξ,y) variables) extended to a complex δ_0-neighborhood of the torus $T_0^n = \mathrm{T}^n\times\{0\}\times\{0\}$ in $\mathcal{Y}_d^c$ (which we denote as $Q^c = Q^c(d)$), where $d\geq d_1/2$. Below we drop the dependence of these functions, as well as the frequencies λ_j, on ε_0. All the estimates are uniform in ε_0.

The extended function H_0 satisfies the estimates

$$|H_0|^{Q^c,\Theta_0} + \|\nabla_y H_0\|_{d-d_H}^{Q^c,\Theta_0} \leq K_1\,, \tag{1.7}$$

where

$$d_H\leq 0\,, \quad d_H<d_1-1\,, \quad d_1\geq 1\,; \tag{1.8}$$

the function H^3 is $O(\|y\|_d^3+|\xi|^2+\|y\|_d\,|\xi|)$. More explicitly, for $\mathfrak{h}=(q,\xi,y)\in Q^c(d)$ the function meets the estimates:

$$\left|H^3(\mathfrak{h};\cdot)\right|^{\Theta_0,\mathrm{Lip}} \leq K_1\left(|\xi|^2+|\xi|\,\|y\|_d+\|y\|_d^3\right), \tag{1.9}$$

$$\left\|\nabla_y H^3(\mathfrak{h};\cdot)\right\|_{d-d_H}^{\Theta_0,\mathrm{Lip}} \leq K_1\left(|\xi|+\|y\|_d^2\right). \tag{1.10}$$

Now equations (1.4) take the form

$$\begin{aligned}\dot q_j &= \omega_j + \frac{\partial}{\partial\xi_j}(\varepsilon_0 H_0 + H^3), \quad \dot\xi_j = -\frac{\partial}{\partial q_j}(\varepsilon_0 H_0+H^3)\,,\\ \dot y &= J\big(A(\theta)y+\nabla_y(\varepsilon_0 H_0+H^3)\big)\,.\end{aligned} \tag{1.11}$$

We remind that by Σ^0 it is denoted the map $\mathrm{T}^n\times\Theta_0\to\mathcal{Y}$, $(q;\theta)\longmapsto(q,0,0)$.

Theorem 1.1. Let the assumptions (1.7) – (1.10) hold together with

1) *spectral asymptotics:*

$$\left|\lambda_j(\theta) - K_2^0 j^{d_1} - K_2^1 j^{d_{1,1}} - \dots - K_2^{r-1} j^{d_{1,r-1}}\right| \le K_1 j^{d_{1,r}} \tag{1.12}$$

with some exponents $d_{1,1}, \dots, d_{1,r}$ such that $d_1 > d_{1,1} > \dots > d_{1,r}$, $d_1 - 1 > d_{1,r}$ and coefficients K_2^ℓ such that

$$K_2^0 \ge K_1^{-1}, \quad \left|K_2^\ell\right| \le K_1 \quad \forall \ell,$$

and

$$\mathrm{Lip}(\lambda_j : \Theta_0 \longrightarrow \mathbb{R}) \le K_1 j^{d_{1,r}} . \tag{1.13}$$

Then there exist integers j_1, M_1 such that if the

2) *nonresonance assumption*

$$|s \cdot \omega + \ell_1 \lambda_1(\theta) + \dots + \ell_{j_1} \lambda_{j_1}(\theta)| \ge \delta_a K_3 > 0 \tag{1.14}$$

holds for all $\theta = (\omega, I)$, all integer n-vectors s and j_1-vectors ℓ such that

$$|s| \le M_1, \quad 1 \le |\ell_1| + \dots + |\ell_{j_1}| \le 2 , \tag{1.15}$$

then for arbitrary $\gamma > 0$ and for small enough positive $\varepsilon_0 = \varepsilon_0(\gamma)$ there exists a Borel subset $\Theta \subset \Theta_0$ and analytic embeddings

$$\Sigma_\theta : \mathrm{T}^n \longrightarrow \mathcal{Y}_{d_c}, \quad \theta \in \Theta, \quad d_c = d - d_H$$

with the following properties:

a) for all I in $\mathcal{J}$

$$\mathrm{mes}(\Omega_0 \backslash \Theta[I]) \le \gamma \,\mathrm{mes}\, \Omega_0 ; \tag{1.16}$$

b) the map

$$\Sigma : \mathrm{T}^n \times \Theta \longrightarrow \mathcal{Y}_{d_c}, \quad (q, \theta) \longmapsto \Sigma_\theta(q)$$

is Lipschitz-close to the map Σ^0. That is, for every $\rho < 1/3$ there exists $C = C(\rho, \gamma)$ and $\varepsilon_0(\rho, \gamma)$ such that

$$\mathrm{dist}_{d_c}(\Sigma^0, \Sigma) \le C \varepsilon_0^\rho, \quad \mathrm{Lip}(\Sigma^0 - \Sigma) \le C \varepsilon_0^\rho \tag{1.17}$$

if $\varepsilon_0 \le \varepsilon_0(\rho, \gamma)$;

c) every torus $\Sigma_\theta(\mathrm{T}^n)$ is invariant for the equations (1.4) and is filled with weak in $\mathcal{Y}_d$ quasiperiodic solutions of the form $z^{\varepsilon_0}(t) = \Sigma_\theta(q + \omega' t)$, where $q \in \mathrm{T}^n$ and the frequency vector $\omega' = \omega'(\theta, \varepsilon_0) \in \mathbb{R}^n$ is close to ω:

$$|\omega - \omega'|^{\Theta_0, \mathrm{Lip}} \le C_1(\gamma) \varepsilon_0^{1/3} ; \tag{1.18}$$

d) all Lyapunov exponents of the solutions z^{ε_0} are equal to zero.

Refinement 1. The numbers j_1, M_1 depend on K, K_1, K_2^ℓ, d, d_1, $d_{1,\ell}$, d_H and n; the numbers $\varepsilon_0(\gamma)$, $\varepsilon_0(\rho,\gamma)$ depend on the same quantities and K_3.

For a proof see Remark at the end of §4.

Refinement 2. The set Θ may be constructed in such a way that the estimates (1.17), (1.18) hold with the exponents ρ and 1/3 replaced by 1.

This statement is proven below in §6.

The unperturbed equations $(1.11)|_{\varepsilon=0}$ have the form

$$\dot{q} = \omega\,, \quad \dot{\xi} = 0\,, \quad \dot{y} = JA(\theta)y\,.$$

In view of (1.6) the operator $JA(\theta)$ has pure imaginary spectrum $\{\pm i\lambda_j(\theta) | j \geq 1\}$. So the invariant torus $T_0^n = \mathrm{T}^n \times \{0\} \times \{0\}$ of this system of equations is elliptic. However, statements a) – c) of Theorem 1.1 also remain true if the torus T_0^n is partially hyperbolic:

Refinement 3. Let all the assumptions of Theorem 1.1 hold, except the first assumption in (1.6),

$$A(\theta)\varphi_j^\pm = \lambda_j(\theta)\varphi_j^\pm\,,$$

which now holds only for $j > j_*$ with some $j_* \geq 0$. Let us also suppose that for $j = 1, \dots, j_* - 1$ the operator $JA(\theta)$ has eigenvectors $W_j^\pm$ (possibly, complex), corresponding to hyperbolic eigenvalues $\pm i\lambda_j(\theta) \notin i\mathrm{R}$, such that the vectors $W_1^\pm, \dots, W_{j_*}^\pm$ span the space $\mathrm{C}\varphi_1^+ + \mathrm{C}\varphi_1^- + \dots + \mathrm{C}\varphi_{j_*}^+ + \mathrm{C}\varphi_{j_*}^-$. Then statements a) – c) of Theorem 1.1 and statements of Refinements 1,2 remain true. The quasiperiodic solutions $z^{\varepsilon_0}(t)$ are partially hyperbolic with $2j_*$ hyperbolic directions.

The proof of this refinement repeat the proof of Theorem 1.1 we give below in §2. The only difference appears during solving the homological equation (Step 4 of the proof). To solve these equations we decompose vectors of the space Y in the complex basis $\{W_j^\pm\}$ of the eigenvectors of the operator $JA(\theta)$. In the elliptic situation $W_j^\pm = (\varphi_j^+ \pm i\varphi_j^-)/\sqrt{2}$ for all $j \geq 1$, but in the partially hyperbolic case the first $2j_*$ vectors $W_j^\pm$, $1 \leq j \leq j_*$, are different. So the same proof requires more complicated notations.

In what follows we restrict ourselves to the less cumbersome elliptic situation only.

Remarks. 1) If assumption 3) of Theorem 2.1.1 (i.e., the nonresonance at the point ω_0) holds, then the assumption 2) of Theorem 1.1 holds with K_3 replaced by $K_3/2$, provided that δ_a is small enough. So Theorem 1.1 jointly with Refinement 2 indeed implies Theorem 2.1.1.

2) Statements of the theorem remain true if the frequencies λ_j depend on ω only and for some resonance functions

$$\Lambda(s,\ell)(\omega) := s \cdot \omega + \ell_1\lambda_1(\omega) + \dots + \ell_{j_1}\lambda_{j_1}(\omega)$$

with (s,ℓ) as in (1.15) instead of (1.14) we have

$$|\nabla_\omega \Lambda(s,\ell)| \geq K_3 \,. \tag{1.19}$$

Indeed, we can delete from Ω_0 the points ω, where $|\Lambda(s,\ell)| \leq \delta_a \gamma C^{-1}$ for all (s,ℓ) as in (1.15), (1.19), and denote the new set Ω_0'. For C large enough $\text{mes}(\Omega_0 \backslash \Omega_0') \leq (\gamma/2)\text{mes}\,\Omega_0$ and relation (1.14) holds for all ω with (s,ℓ) as in (1.15) and K_3 replaced by $\min(K_3, \gamma C^{-1})$. So the theorem can be applied to Ω_0 replaced by Ω_0'.

3) In fact, the invariant tori $\Sigma_\theta(\mathrm{T}^n)$ of the equations (1.4) are smoother than it was stated in the theorem: the assertions of the theorem remain true with d_c replaced by

$$d_c = d - d_H + d_1 - 1 \,.$$

For the proof see [K6] and the discussion in the remark following Lemma 2.2 in §2 below.

4) The quasiperiodic solutions $z^{\varepsilon_0}(t)$, constructed in Theorem 1.1, possess additional regularity properties: the equations (1.11), linearized above the solution $z^{\varepsilon_0}(t)$, can be reduced to constant-coefficient linear equations by means of a time-quasiperiodic linear change of variables. See §7 below.

5) The map Σ can be analytically in q extended to the complex $\delta_0/2$-neighborhood of the n-torus, where the estimates (1.17) remain true.

For the applications to small-amplitude solutions of nonlinear partial differential equations, mentioned in Part 3.2.B of Introduction, we need a version of Theorem 1.1 for equations with the frequency vector ω varying in a small domain (i.e., with the radius δ_a in (1.2) of order ε in some positive degree).

The corresponding result is stated below in Theorem 1.2, where for a Banach space B, a metric space M and a map $G: M \to B$ we denote by $[\cdot]_B^{M,\mathrm{Lip}}$ the weighted Lipschitz norm

$$[G]_B^{M,\mathrm{Lip}} = \max\{\sup |G(m)|_B \;,\; \delta_a \,\mathrm{Lip}\, G\} \,. \tag{1.20}$$

Observe that this norm is equivalent to the norm defined in (2.0.1) if δ_a is of order one (as in Theorem 1.1), and that

$$[fG]_B^{M,\mathrm{Lip}} \leq 2[f]^{M,\mathrm{Lip}}[G]_B^{M,\mathrm{Lip}} \tag{1.21}$$

for a scalar function f.

To state the result we should suppose that the perturbation $\varepsilon_0 H^0$ of the integrable hamiltonian $\omega \cdot \xi + \frac{1}{2}\langle A(\theta)y, y\rangle$ is of a less general form than in (1.7) – (1.10). We assume that

$$\varepsilon_0 H^0 = \varepsilon_0 H_0 + \delta_a H_0^3 \,,$$

(i.e., in (1.11) $H^3 = \delta_a H_0^3$) and

$$[H_0]^{Q^c,\Theta_0} + [\nabla_y H_0]_{d_c}^{Q^c,\Theta_0} \leq K_1$$

and for $\mathfrak{h} \in Q^c$

$$\begin{aligned} \left[H_0^3(\mathfrak{h};\cdot)\right]^{\Theta_0,\mathrm{Lip}} &\leq K_1\left(|\xi|^2 + |\xi|\,\|y\|_d^2 + \|y\|_d^3\right) , \\ \left[\nabla_y H_0^3(\mathfrak{h};\cdot)\right]_{d_c}^{\Theta_0,\mathrm{Lip}} &\leq K_1\left(|\xi|\,\|y\|_d + \|y\|_d^2\right) . \end{aligned} \tag{1.22}$$

Let us fix some $0 < \mu \le 1$.

Theorem 1.2. For each $\gamma > 0$, for all small enough ε_0 (i.e., for $\varepsilon_0 < \varepsilon_0(\gamma)$) and for δ_a such that

$$\delta_a > \varepsilon_0^{1-\mu}/C \tag{1.23}$$

there exist a Borel subset $\Theta \subset \Theta_0$ and analytic embeddings Σ_θ, $\theta \in \Theta$, such that the statements a) – d) of Theorem 1.1 hold with the estimates (1.17), (1.18) refined as follows:

$$[\Sigma^0 - \Sigma]_{d_c}^{\mathrm{T}^n \times \Theta, \mathrm{Lip}} \le C\varepsilon_0^{\mu}, \tag{1.17'}$$

$$[\omega - \omega']^{\Theta, \mathrm{Lip}} \le C\varepsilon_0 . \tag{1.18'}$$

Remarks. 1) By the Refinement 1 to Theorem 1.1 the numbers j_1, M_1 in the assumption 2) of Theorem 1.1 do not depend on δ_a (and so remain bounded when δ_a tends to zero with ε_0 as in (1.23)).

2) The statements of Remarks 2, 5 to Theorem 1.1 remain true with the same proof.

2. Proof of Theorem 1.1

2.1. Notations and a scheme of the proof.

For the Hilbert scale $\{Y_s\}$ we denote by Y_s^c complexification of the space Y_s and extend the scalar product $\langle\cdot,\cdot\rangle = \langle\cdot,\cdot\rangle_Y$ to complex-bilinear pairing $Y_{-s}^c \times Y_s^c \to \mathrm{C}$. We denote by $\mathcal{L}^s(Y_a^c, Y_b^c)$ the space of continuous linear operators from Y_a^c to Y_b^c, symmetric with respect to the pairing $\langle\cdot,\cdot\rangle$ (that is, $\langle Ly_1, y_2\rangle = \langle y_1, Ly_2\rangle$ for all y_1, y_2 from Y_∞^c).

We denote by $U(\delta)$ the complex δ-neighborhood of the n-torus:

$$U(\delta) = \{\xi \in \mathrm{C}^n/2\pi \mathrm{Z}^n \mid |Im\,\xi| < \delta\},$$

denote $\zeta = 2(1^{-2} + 2^{-2} + \cdots)$, and denote by $\{e_m\}$ the numbers

$$e_m = \begin{cases} 0, & m = 0 \\ (1^{-2} + \cdots + m^{-2})/\zeta, & m \ge 1 \end{cases}$$

(thus $e_m < 1/2$ for all m). We shall use two decreasing sequences of parameters $\{\varepsilon_m\}$ and $\{\delta_m\}$:

$$\varepsilon_m = \varepsilon_0^{(1+\rho)^m}, \quad \delta_m = \delta_0(1 - e(m))$$

(thus $\delta_m > \frac{1}{2}\delta_0$ for all m) and shall consider analytic functions of argument $(q, \xi, y) \in \mathcal{Y}$ with radius of analyticity δ_m in q, $\varepsilon_m^{2/3}$ in ξ and $\varepsilon_m^{1/3}$ in y. Thus we consider the following domains of analyticity:

$$O_m^c = U_m \times O(\varepsilon_m^{2/3}, \mathrm{C}^n) \times O(\varepsilon_m^{1/3}, Y_d^c), \quad \text{where} \quad U_m = U(\delta_m).$$

Domains O_m^c form neighborhoods in $\mathcal{Y}_d^c$ of the torus

$$T_0^n = \mathrm{T}^n \times \{0\} \times \{0\} \subset \mathcal{Y}_d .$$

The real part of domain U_m is n-torus T^n and the real part of O_m^c is O_m, where

$$O_m = \mathrm{T}^n \times O(\varepsilon_m^{2/3}, \mathrm{R}^n) \times O(\varepsilon_m^{1/3}, Y_d) .$$

These domains form decreasing families with m increasing:

$$U_0 \supset U_1 \supset \cdots \supset \mathrm{T}^n , \quad O_0^c \supset O_1^c \supset \cdots \supset U\Big(\frac{1}{2}\delta_0\Big) \times \{0\} \times \{0\} \supset T_0^n ,$$

$$O_0 \supset O_1 \supset \cdots \supset T_0^n .$$

Let the set $\Theta_0 = \Omega_0 \times \mathcal{J}$ be the same as above. For $m = 1,2,\ldots$ we will construct Borel subsets Θ_m of Θ_0 such that

$$\operatorname{mes}(\Omega_0 \backslash \Theta_m[I]) \le K^{-1}\gamma\delta_a^n e(m) \tag{2.1}$$

for all I (for $m = 0$ this estimate trivially holds). We omit dependence of functions and sets on the parameter ε_0. All estimates will be uniform with respect to ε_0.

On domain O_m^c we consider the hamiltonian $\mathcal{H}_m$, depending on the parameter $\theta \in \Theta_m$

$$\begin{aligned} \mathcal{H}_m &= H_{0m}(\xi, y; \theta) + \varepsilon_m H_m(q, \xi, y; \theta) + H^3(q, \xi, y; \theta) , \\ H_{0m} &= \xi \cdot \Lambda_m(\theta) + \frac{1}{2}\langle A_m(\theta) y, y\rangle . \end{aligned} \tag{2.2}$$

Here the function H^3 is the same as in the equations (1.11) and the vector Λ_m is close to ω:

$$|\Lambda_m(\omega, I) - \omega|^{\Theta_m,\mathrm{Lip}} \le C\varepsilon_0^{\tilde\rho} e(m) ; \tag{2.3}$$

during the proof of the theorem in §§2–3

$$\tilde\rho = \rho \tag{2.4}$$

(later we shall improve the estimate (2.3) and increase $\tilde\rho$ to prove Refinement 2 and Theorem 1.2). The operator $A_m(\theta)$ is diagonal in the basis $\{\varphi_j^\pm\}$ and for all j

$$\big(A_m(\theta) - A(\theta)\big)\varphi_j^\pm = \beta_{jm}(\theta)\varphi_j^\pm , \tag{2.5}$$

where

$$|\beta_{jm}|^{\Theta_m,\mathrm{Lip}} \le C\varepsilon_0^{\tilde\rho} e(m) j^{d_H} . \tag{2.6}$$

We suppose that the function H_m in (2.2) is analytic in O_m^c, real for real arguments and

$$|H_m|^{O_m^c;\Theta_m} \le C_*(m) \equiv K_4^{m+1} , \tag{2.7}$$

$$\|\nabla_y H_m\|_{d_c}^{O_m^c,\Theta_m} \le \varepsilon_m^{-1/3} C_*(m) , \quad d_c = d - d_H . \tag{2.8}$$

For $m = 0$ the hamiltonian $\mathcal{H}_0$ in (1.3) has the form (2.2) with $\Lambda_0(\omega, I) \equiv \omega$, $A_0 = A$ and the estimates (2.7), (2.8) are fulfilled by the theorem's assumptions.

Hamiltonian equations with the hamiltonian $\mathcal{H}_m$ have the form

$$\dot{q} = \Lambda_m(\theta) + \nabla_\xi(\varepsilon_m H_m + H^3)(q,\xi,y;\theta) , \tag{2.9}$$

$$\dot{\xi} = -\nabla_q(\varepsilon_m H_m + H^3)(q,\xi,y;\theta) , \tag{2.10}$$

$$\dot{y} = J\big(A_m(\theta)y + \nabla_y(\varepsilon_m H_m + H^3)(q,\xi,y;\theta)\big) , \tag{2.11}$$

and for $m = 0$ coincide with the initial equations (1.11).

We remark that by (1.9), (1.10) the function $\varepsilon_m^{-1}H^3$ also satisfies estimates (2.7), (2.8) in the domain O_m^c. So we can denote $\varepsilon_m H_m + H^3 = \varepsilon_m \tilde{H}_m$; where the function $\tilde{H}_m$ meets the estimates (2.7), (2.8) with some larger K_4, and obtain for hamiltonian $\mathcal{H}_m$ a representation $\mathcal{H}_m = H_{0m} + \varepsilon_m \tilde{H}_m$ (i.e., without the term H^3). We do not do it because the representation (2.2) will allow us below in §§6, 8 improve the estimates of Theorem 1.1 for the set Θ and the maps Σ_θ, and thus prove Refinement 2 and Theorem 1.2.

The theorem will be proved via the KAM-procedure. For $m = 0,1,2,\dots$ we shall construct a canonical transformation $S_m : O_{m+1} \to O_m$ which is well-defined, provided that $\theta \in \Theta_{m+1}$, and transforms the equations (2.9) – (2.11) into Hamiltonian equations in O_{m+1} with the hamiltonian of the form (2.2) with $m := m+1$. For $\theta \in \Theta = \cap\Theta_m$ the limit transformation $\Sigma : S_0 \circ S_1 \circ \cdots$ transforms equations (1.11) into an equation in the set $\cap O_m = T_0^n$. The last one has the solutions $\big(q + t\Lambda_\infty(\theta), 0, 0\big)$, where $\Lambda_\infty = \lim \Lambda_m$ and $q \in T^n$. So for $\theta \in \Theta$ equation (1.11) has the desired quasiperiodic solutions of the form $\Sigma(q + t\omega', 0, 0)$, where $\omega' = \Lambda_\infty$.

The proof goes in five steps.

At *Step 1* we split the perturbation $\varepsilon_m H_m$ into an "essential" part $\varepsilon_m H'_{2m}$ which is linear in ξ, quadratic in y, and an unessential part

$$\varepsilon_m H_3 = \mathcal{O}^3(q,\xi,y) \overset{\text{def}}{=} O\big(|\xi|^2 + \|y\|_d^3 + |\xi|\,\|y\|_d\big) .$$

Step 2 is an averaging. We extract from the essential part $\varepsilon_m H'_{2m}$ of the perturbation the "averaged terms" $\varepsilon_m\big(\xi\cdot h^{0\xi} + \langle y, h^{0yy}y\rangle\big)$, where

$$h^{0\xi} = (2\pi)^{-n}\int_{T^n} \nabla_\xi H'_{2m}(q,\xi,0)|_{\xi=0}\,dq ,$$

and h^{0yy} is the diagonal (in the complex basis $\{(\varphi_j^+ \pm i\varphi_j^-)/\sqrt{2}\}$) part of the averaged y-Hessian of H'_{2m}. That is, the diagonal part of the operator

$$(2\pi)^{-n}\int_{T^n} \frac{\partial^2}{\partial y^2} H'_{2m}(q,0,y)|_{y=0}\,dq .$$

We add these terms to H_{0m} and thus obtain the integrable part H_{0m+1} of the next iteration. The remaining part of $\varepsilon_m H'_{2m}$ is denoted $\varepsilon_m H_{2m}$. So

$$\mathcal{H}_m = H_{0m+1} + \varepsilon_m H_{2m} + \mathcal{O}^3(q,\xi,y) .$$

At *Step 3* we formally construct the transformation S_m as the time-one shift along trajectories of the Hamiltonian vector field with some ξ-linear, y-quadratic hamiltonian $\varepsilon_m F$,

$$F = f^q(q) + \xi \cdot f^\xi(q) + \langle y, f^y(q)\rangle + \langle y, f^{yy}(q)y\rangle \, .$$

We formally apply Proposition 1.4.3 and write down the transformed hamiltonian as

$$\begin{aligned} \mathcal{H}_m \circ S_m &= \mathcal{H}_m + \varepsilon_m\{F, \mathcal{H}_m\} + O(\varepsilon_m^2) \\ &= H_{0m+1} + \varepsilon_m\big(H_{2m} + \{F, H_{0m+1}\}\big) \\ &+ \big(\varepsilon_m\{F, \mathcal{O}^3\} + \mathcal{O}^3 + O(\varepsilon_m^2)\big) \, . \end{aligned} \tag{2.12}$$

We want to construct the hamiltonian F in such a way that the second term in the r.h.s. of (2.12) vanish. For this end the functions $f^q, \ldots, f^{yy}$ should satisfy some linear homological equations (see equations (2.31) – (2.33) below). The most complicated among the equations is the equation for the operator-valued function f^{yy},

$$(\Lambda_{m+1} \cdot \nabla_q) f^{yy} + [f^{yy}, JA_{m+1}] = h^{yy} \, , \tag{2.13}$$

where $h^{yy} = \frac{\partial^2}{\partial y^2} H_{2m}(q, 0, y)|_{y=0}$.

At *Step 4* we solve the homological equations and prove that the hamiltonian $\mathcal{H}_m \circ S_m$ is of the form (2.2) with $m = m+1$.

The homological equations are solved for θ outside a set $\Delta\Theta$ which is defined as a set of parameters θ, where one of the denominators in the formulae giving solutions of the homological equations is very small. Here the key is an estimate for measure of the set $\Delta\Theta$, proven in §4.

In particular, to estimate the denominators corresponding to equation (2.13) we observe that the (j,k)-matrix element $f_{jk}(q)$ of the operator f^{yy} is given by Fourier series

$$f_{jk}(q) = \sum_{s \in \mathbf{Z}^n} \hat{f}_{jk}(s) e^{is\cdot q} \, , \tag{2.14}$$

where the expression for $\hat{f}_{jk}(s)$ has the denominator D,

$$D = \Lambda_{m+1} \cdot s \pm \lambda'_j \pm \lambda'_k \, ,$$

and $\{\lambda'_j = \lambda'_j(\theta)\}$ are the eigenvalues of the operator $A_{m+1}(\theta)$ (close to the eigenvalues of $A(\theta)$ by (2.5), (2.6)). The most complicated for an investigation are the denominators with the opposite signs for λ'_j and λ'_k. To handle them we should estimate the measure of the set

$$\{\theta \mid |D| = \left|\Lambda_{m+1} \cdot s + \lambda'_j - \lambda'_k\right| \quad \text{is small for some } s, \, j, \, k\} \, .$$

To do it we find some $C > 1$ such that the numbers $\Lambda_{m+1} \cdot s$ lie in the segment $\Delta_s = [-C\,|s|\,, C\,|s|]$. As λ'_j grows as j^{d_1} when $j \to \infty$, then for $\lambda'_{jk} = \lambda'_j - \lambda'_k$ we have an estimate $\left|\lambda'_{jk}\right| \geq C^{-1} \min(|j|, |k|)^{d_1 - 1}$. So if $d_1 > 1$, then only a finite system of these numbers hit to the segment Δ_s. We define "bad subsets" Θ^s, $s \in \mathbf{Z}^n$, as the sets of parameters θ such that the number $\Lambda_{m+1} \cdot s$ hits to γ_s-neighborhood of the discrete set $M_s = \{\lambda'_{jk}\} \bigcap \Delta_s$, and define the "bad set" $\Delta\Theta$ as the union $\bigcup_{s \in Z^n} \Theta^s$. If the numbers γ_s decrease fast enough with $|s|$ growing, then the measure of $\Delta\Theta$ is small and outside $\Delta\Theta$ the estimate $|D| > \gamma_s$ holds. We construct the numbers γ_s diminishing "not too fast" and show that out of $\Delta\Theta$ the series (2.14) define an analytic solution of (2.13).

If $d_1 = 1$, then the set M_s is infinite. Generically the closure of M_s has positive Lebesgue measure and the scheme we have just explained does not work. To save the scheme we make use of the higher order terms of the spectral asymptotics (1.12) and observe that

$$\left|\lambda'_{jk} - K_2^0(k-j)\right| \le C\,|k-j| \min(k,j)^{-\kappa}$$

for all k, j with some $\kappa > 0$. So the only limiting points of the infinite set M_s are points of the one-dimensional lattice $\Gamma_s = \{\ell K_2^0 \mid \ell \in Z\} \bigcap \Delta_s$. Now we define "the first bad set" $\Delta_1\Theta$ as the set of parameters θ such that $\Lambda_{m+1} \cdot s$ hits to γ_s-neighborhood of Γ_s for some s, and define "the second bad set" $\Delta_2\Theta$ essentially as for $d_1 > 1$, but taking into account only the points λ'_{jk}, remote from $\Gamma_\infty = \bigcup \Gamma_s$ (in each segment Δ_s there are only finitely many of them). Now the measure of the set $\Delta\Theta = \Delta_1\Theta \bigcup \Delta_2\Theta$ can be estimated and it can be shown that with a suitable choice of the numbers γ_s the series converge to an analytic solution.

After the homological equations are solved for $\theta \in \Theta_{m+1} = \Theta_m \backslash \Delta\Theta$, we study smoothness of the flow of the Hamiltonian vector-field with the hamiltonian F and prove that the third term in the r.h.s. of (2.12) is $\varepsilon_{m+1} H_{m+1} + H^3$ with some function H_{m+1} satisfying (2.7), (2.8) with $m = m+1$.

So for θ in Θ_{m+1} the transformed hamiltonian $\mathcal{H}_{m+1} = \mathcal{H}_m \circ S_m$ again has the form (2.2) and we can iterate the procedure infinitely.

At the last *Step 5* we prove that for θ in $\Theta = \bigcap \Theta_m$ the limit transformation $\Sigma = S_0 \circ S_1 \circ \cdots$ exists and integrate equations (1.11).

2.2. The Proof

We shall use intermediate domains between O_m^c and O_{m+1}^c. For this end we define the intermediate numbers

$$\delta_m^j = \frac{6-j}{6}\delta_m + \frac{j}{6}\delta_{m+1}\,, \quad 0 \le j \le 5 \tag{2.15}$$

(so $\delta_m = \delta_m^0 > \delta_m^1 > \cdots > \delta_m^5$), and denote

$$O_m^{jc} = U(\delta_m^j) \times O((2^{-j}\varepsilon_m)^{2/3}, \mathbf{C}^n) \times O((2^{-j}\varepsilon_m)^{1/3}, Y_d^c)\,,$$

$$U_m^j = U(\delta_m^j)\,.$$

If $\varepsilon_0 << 1$, then $2^{-j}\varepsilon_m > \varepsilon_{m+1}$ for $j = 1, \ldots, 5$, and the domains O_m^{jc} are neighborhoods of O_{m+1}^c in $O_m^c : O_m^c \supset O_m^{1c} \supset \cdots \supset O_m^{5c} \supset O_{m+1}^c$.

We denote by $C, C_1, C_2, \ldots$ different positive constants independent of ε_0 and m; by $C(m)$, $C_1(m), \ldots$ different functions of m of the form $C(m) = C_1 m^C$. By

$C_*, C_{*1}, \ldots, C_*(m), C_{*1}(m), \ldots$ we denote fixed constants and fixed functions of the form $C(m)$. The constants C depend on the radius δ_a, unless otherwise is stated. Observe that for each $C(m)$ and each $\sigma < 0$ the estimate $C(m) < \varepsilon_m^\sigma$ holds for all m, provided that ε_0 is small enough.

Step 1. Splitting the perturbation.

We extract from H_m the linear in ξ and quadratic in y part:

$$\begin{aligned} H_m(q,\xi,y;\theta) = h^q(q;\theta) + \xi \cdot h^{1\xi}(q;\theta) + \\ + \langle y, h^y(q;\theta)\rangle + \langle y, h^{yy}(q;\theta)y\rangle + H_{3m}(q,\xi,y;\theta) , \end{aligned} \tag{2.16}$$

where

$$H_{3m} = O\big(|\xi|^2 + \|y\|_d^3 + |\xi|\,\|y\|_d\big) , \tag{2.17}$$

h^q is a scalar, $h^{1\xi}$ is an n-vector, $h^y \in Y$ and h^{yy} is a selfadjoint operator in Y.

Step 2. Averaging.

We may change H_m on a θ-depending constant, and so may suppose that

$$\int h^q(q;\theta)dq/(2\pi)^n = 0 .$$

Here and in what follows

$$\int f(q)dq/(2\pi)^n = (2\pi)^{-n}\int_{\mathrm{T}^n} f(q)dq$$

for an arbitrary vector-valued function integrable on T^n.

Now we average the essential part of the perturbation H_m. That is, we extract from $\xi \cdot h^{1\xi}$ and $\langle y, h^{yy}y\rangle$ two integrable terms. The first one is equal to $\xi \cdot h^{0\xi}(\theta)$, where

$$h^{0\xi} = \int h^{1\xi}(q;\theta)dq/(2\pi)^n .$$

The second one is $\langle h^{0yy}(\theta)y, y\rangle$, where the operator $h^{0yy}(\theta)$ is diagonal in the basis $\{\varphi_j^\pm\}$,

$$h^{0yy}(\theta)\varphi_j^\pm = b_j(\theta)\varphi_j^\pm ,$$

and

$$b_j(\theta) = \frac{1}{2}\int \big(\langle h^{1yy}(q;\theta)\varphi_j^+, \varphi_j^+\rangle + \langle h^{1yy}(q;\theta)\varphi_j^-, \varphi_j^-\rangle\big)dq/(2\pi)^n .$$

We denote

$$h^\xi = h^{1\xi} - h^{0\xi}, \quad h^{yy} = h^{1yy} - h^{0yy} \tag{2.18}$$

and rearrange the terms in (2.2) as follows:

$$\mathcal{H}_m = H_{0m+1}(\xi, y;\theta) + \varepsilon_m(H_{2m} + H_{3m}) + H^3 , \tag{2.19}$$

where

$$H_{0m+1} = \xi \cdot \Lambda_{m+1} + \frac{1}{2}\langle A_{m+1}y, y\rangle$$

with

$$\Lambda_{m+1} = \Lambda_m(\theta) + h^{0\xi}(\theta), \quad A_{m+1} = A_m(\theta) + 2\varepsilon_m h^{0yy}(\theta) ;$$

the function H_{3m} is the same as in (2.16) and

$$H_{2m} = h^q + \xi \cdot h^\xi + \langle y, h^y \rangle + \langle y, h^{yy} y \rangle . \tag{2.20}$$

It is clear that after the transformation (2.18) the vector-function h^ξ has zero mean-value. In the complex eigen-basis $\{\frac{1}{\sqrt{2}}(\varphi_j^+ \pm i\varphi_j^-)\}$ of the operator JA the diagonal elements of the matrix of the operator h^{1yy} are equal to $\frac{1}{2}(\langle h^{1yy}\varphi_j^+, \varphi_j^+ \rangle + \langle h^{1yy}\varphi_j^-, \varphi_j^- \rangle)$ and diagonal elements of the matrix of the operator h^{0yy} are equal to $b_j(\theta)$. So the diagonal elements of the matrix of the operator $h^{yy} = h^{1yy} - h^{0yy}$ have zero mean-value. These properties of hamiltonian (2.20) will be critical to solve homological equations at Step 4.

Now we estimate the terms of the decompositions (2.16) and (2.19), (2.20).

Lemma 2.1. If $\varepsilon_0 << 1$ then the terms of the decomposition (2.16) may be estimated as follows:

a)

$$|h^q|^{U_m,\Theta_m} \le C_*(m) , \tag{2.21}$$

$$\left|h^{1\xi}\right|^{U_m,\Theta_m} \le C_*(m)\varepsilon_m^{-2/3} , \tag{2.22}$$

$$\|h^y\|_{d_c}^{U_m,\Theta_m} \le C_*(m)\varepsilon_m^{-1/3} , \tag{2.23}$$

($d_c = d - d_H$ and the number $C_*(m)$ is the same as in (2.7), (2.8));

b) the operator h^{yy} is symmetric in Y, is real for real q and

$$\left\|h^{1yy}\right\|_{d,d_c}^{U_m,\Theta_m} \le C_*(m)\varepsilon_m^{-2/3} ; \tag{2.24}$$

c) in the domain O_{m+1}^c the term $\varepsilon_m H_{3m}$ is twice smaller than the admissible disparity $\varepsilon_{m+1}H_{m+1}$ of the next step:

$$\varepsilon_m \left|H_{3m}\right|^{O_{m+1}^c,\Theta_m} \le \frac{1}{2}C_*(m+1)\varepsilon_{m+1} ,$$

$$\varepsilon_m \left\|\nabla_y H_{3m}\right\|_{d_c}^{O_{m+1}^c,\Theta_m} \le \frac{1}{2}C_*(m+1)\varepsilon_{m+1}^{2/3} ,$$

provided that K_4 in (2.7) is large enough;

d) the functions H_{2m}, H_{3m} are analytic in (q,ξ,y) and are real for real arguments.

Proof. a) The estimate (2.21) results from (2.7) because $h^q(q;\theta) = H_m(q,0,0;\theta)$. To prove (2.22) we observe that $h^{1\xi}(q;\theta) = \nabla_\xi H_m(q,0,0;\theta)$. So the estimate for $\left|h^{1\xi}\right|$ results by application Cauchy estimate to the map $\xi \longmapsto H_m(q,\xi,0;\theta)$ (see below Theorem D in Appendix D with $O = O(\varepsilon_m^{2/3}, \mathbf{C}^n)$ and $w = 0$). To get the estimate for the Lipschitz constant in θ one should apply Cauchy estimate to the map $\xi \longmapsto H_m(q,\xi,0;\theta_1) - H_m(q,\xi,0;\theta_2)$ and argue as above.

The estimate (2.23) results from (2.8) with $y=0$.

b) The estimate (2.24) results by applying Cauchy estimate to the map $\nabla_y H_m : y \longmapsto \nabla_y H_m(q,0,y;\theta)$, because $h^{yy} = \frac{1}{2}(\nabla_y H_m(0))_*$. The inclusion $h^{yy} \in \mathcal{L}^s(Y_d;Y_{d_c})$ results from the general fact that the Hessian of a function is a symmetric linear operator.

c) Let $\mathfrak{h} = (q,\xi,y) \in O^c_{m+1}$ and $\nu = \varepsilon_m^{\rho/3}$. Then $(q,(z/\nu)^2\xi,(z/\nu)y) \in O^c_m$ for $z \in \mathbf{C}$, $|z| \le 1$. Let us consider the function $z \longmapsto H_m(q,(z/\nu)^2\xi,(z/\nu)y;\theta)$ and its Taylor series at zero:

$$H_m(q,(\frac{z}{\nu})^2\xi,(\frac{z}{\nu})y;\theta) = h_0 + h_1 z + h_2 z^2 + \cdots .$$

By (2.7), $|h_k| \le C_*(m)$ for all k. Since $H_{3m}(\mathfrak{h};\theta) = h_3\nu^3 + h_4\nu^4 + \cdots$, then

$$\varepsilon_m \left| H_{3m}(\mathfrak{h};\theta)\right| = \varepsilon_m \left| h_3\nu^3 + h_4\nu^4 + \cdots \right| \le \frac{C_*(m)\varepsilon_m^{1+\rho}}{1-\varepsilon_m^{\rho/3}} \le \frac{C_*(m+1)\varepsilon_{m+1}}{2},$$

if $K_4 >> 1$. In a similar way one can estimate the Lipschitz constant of H_{3m}.

To estimate $\nabla_y H_{3m}$ we consider the map

$$z \longrightarrow \nabla_y H_m(q,(\frac{z}{\nu})^2\xi,(\frac{z}{\nu})y;\theta) = h'_0 + h'_1 z + \cdots \in Y^c_{d_c} .$$

By (2.8) $\|h'_k\|_{d_c} \le \varepsilon_m^{-1/3} C_*(m)$ for all k. So

$$\varepsilon_m \left\| \nabla_y H_{3m}(\mathfrak{h};\theta)\right\|_{d_c} = \varepsilon_m \left\| h'_2\nu^2 + h'_3\nu^3 + \cdots \right\|_{d_c} \le \frac{\nu^2}{1-\nu}\varepsilon_m^{2/3} C_*(m) \le \frac{1}{2}\varepsilon_{m+1}^{2/3} C_*(m+1) .$$

A similar estimate holds for the Lipschitz constant, so the assertion c) is proved.

d) The analyticity of the functions is evident. Their real-valuedness for real arguments results from the real-valuedness of $\mathcal{H}_m$. □

As a consequence of estimates (2.22), (2.24) we obtain estimates for the terms of the partially averaged hamiltonian (2.20):

Corollary. The vector Λ_{m+1} and the operator A_{m+1} satisfy estimates (2.3) and (2.5), (2.6) with m replaced by $m+1$; the vector h^ξ and the operator h^{yy} meet the estimates

$$|h^\xi|^{U_m,\Theta_m} \le C(m)\varepsilon_m^{-2/3}, \tag{2.25}$$

$$\|h^{yy}\|^{U_m,\Theta_m}_{d,d_c} \le C(m)\varepsilon_m^{-2/3} . \tag{2.26}$$

Proof. Averaging estimates (2.22), (2.24) and taking into account (1.5), we obtain the inequalities

$$|h^{0\xi}|^{\Theta_m,\mathrm{Lip}} \le C_*(m)\varepsilon_m^{-2/3}, \quad |b_j|^{\Theta_m,\mathrm{Lip}} \le C(m)\varepsilon_m^{-2/3} j^{d_H} . \tag{2.27}$$

The first one jointly with the estimate (2.3) for Λ_m implies the estimate for Λ_{m+1} (because $C(m)\varepsilon_m^{1/s} < C\varepsilon_0^{\beta}(m+1)^{-2}$ if $\varepsilon_0 << 1$), and jointly with (2.22) implies (2.25). The second estimate in (2.27) jointly with (2.5), (2.6) implies the relations (2.5), (2.6) for A_{m+1}. Jointly with (2.24) it implies (2.26). □

Step 3. Formal construction of the transformation S_m and derivation the homological equations.

We construct S_m as a time-one shift along the trajectories of an auxiliary Hamiltonian vector-field

$$\dot{q} = \varepsilon_m \nabla_\xi F\,, \qquad \dot{\xi} = -\varepsilon_m \nabla_q F\,, \qquad \dot{y} = \varepsilon_m J \nabla_y F\,, \tag{2.28}$$

where the hamiltonian F is ξ-linear and y-quadratic:

$$F = f^q(q;\theta) + \xi \cdot f^\xi(q;\theta) + \langle y, f^y(q;\theta)\rangle + \langle y, f^{yy}(q;\theta)y\rangle\,.$$

The flow of equations (2.28) consists of canonical transformations $\{S^t\}$ of the phase space (Theorem 1.2.4), and $S_m = S^t|_{t=1}$. We denote points of the space $\mathcal{Y}$ by $(q,\xi,y) = \mathfrak{h}$. Formally applying Proposition 1.4.3 we get:

$$\mathcal{H}_m(S_m(\mathfrak{h};\theta);\theta) = \mathcal{H}_m(\mathfrak{h};\theta) + \varepsilon_m\{F, \mathcal{H}_m\} + O(\varepsilon_m^2)\,.$$

By Lemma 2.1 and assumption (1.8) of the theorem, for $\mathfrak{h} \in O_{m+1}$

$$\left|\varepsilon_m H_{3m}(\mathfrak{h}) + H^3(\mathfrak{h})\right| = O(\varepsilon_m^{1+\rho}) = O(\varepsilon_{m+1})\,.$$

So

$$\begin{aligned}\mathcal{H}_m(S_m(\mathfrak{h})) &= H_{0m+1}(\mathfrak{h}) + \varepsilon_m\Big(H_{2m}(\mathfrak{h}) + \big\{F(\mathfrak{h}), H_{0m+1}(\mathfrak{h})\big\}\Big) + O(\varepsilon_m^{1+\rho})\\ &= H_{0m+1} + \varepsilon_m\big(H_{2m} - \nabla_q F \cdot \nabla_\xi H_{0m+1} + \langle J\nabla_y F, \nabla_y H_{0m+1}\rangle\big) + O(\varepsilon_{m+1})\,.\end{aligned}$$

As $\nabla_\xi H_{0m+1} = \Lambda_{m+1}$ and $\nabla_y H_{0m+1} = A_{m+1}y$, then we may denote

$$\omega' = \Lambda_{m+1}(\omega,\theta)\,, \qquad \frac{\partial}{\partial\omega'} = \omega' \cdot \nabla_q \tag{2.29}$$

and rewrite $\mathcal{H}_m \circ S_m$ as follows:

$$\begin{aligned}\mathcal{H}_m(S_m(\mathfrak{h};\theta);\theta) &= H_{0m+1}\\ &+ \varepsilon_m\Big[-\partial f^q/\partial\omega' - \xi \cdot \partial f^\xi/\partial\omega' - \langle y, \partial f^y/\partial\omega'\rangle - \langle y, (\partial f^{yy}/\partial\omega')y\rangle\\ &+ \langle A_{m+1}y, Jf^y\rangle + 2\langle A_{m+1}y, Jf^{yy}y\rangle + h^q + \xi \cdot h^\xi + \langle y, h^y\rangle + \langle y, h^{yy}y\rangle\Big]\\ &+ O(\varepsilon_{m+1})\end{aligned} \tag{2.30}$$

(the term in the square brackets equals $H_{2m} + \{F, H_{0m+1}\}$).

We try to find the hamiltonian F in such a way that the contents of the square brackets in the r.h.s. of (2.30) vanish. For this end we have to find the functions f^q, f^ξ, f^y and f^{yy}, solving the homological equations:

$$\partial f^q/\partial\omega' = h^q(q;\theta), \quad \partial f^\xi/\partial\omega' = h^\xi(q;\theta) \tag{2.31}$$

$$\partial f^y/\partial\omega' - A_{m+1}(\theta)Jf^y = h^y(q;\theta)\,, \tag{2.32}$$

$$\partial f^{yy}/\partial\omega' + f^{yy}JA_{m+1} - A_{m+1}Jf^{yy} = h^{yy}(q;\theta)\,. \tag{2.33}$$

If the functions $f^q,\ldots,f^{yy}$ satisfy (2.31)–(2.33) and the function F is defined as above, then the contents of the square brackets in the r.h.s. of (2.30) vanishes and

$$\{F, H_{0m+1}\} = -H_{2m}\,. \tag{2.34}$$

Step 4. Solving the homological equations and investigation the transformation S_m.

The following lemma, stating the existence of analytic solutions of the homological equations, is proven below in §3:

Lemma 2.2. If $\varepsilon_0 << 1$, then there exists a Borel subset $\Theta_{m+1} \subset \Theta_m$ such that for all I

$$\text{mes}(\Theta_m \backslash \Theta_{m+1})[I] \le K^{-1}\gamma\delta_a^n(m+1)^{-2} \tag{2.35}$$

and for $\theta \in \Theta_{m+1}$ the homological equations have analytic solutions:

a) equations (2.31) have analytic solutions f^q, f^ξ real for real arguments and such that

$$|f^q|^{U^1_m,\Theta_{m+1}} \le C(m)\,, \quad |f^\xi|^{U^1_m,\Theta_{m+1}} \le \varepsilon_m^{-2/3}C(m)\,; \tag{2.36}$$

b) equation (2.32) has an analytic solution f^y such that

$$\|f^y\|^{U^1_m,\Theta_{m+1}}_{d_c+d_1} \le C(m)\varepsilon_m^{-1/3}\,; \tag{2.37}$$

c) equation (2.33) has a solution $f^{yy} \in \mathcal{A}^R_{\Theta_{m+1}}(U^1_m;\mathcal{L}^S(Y_d,Y_{d_c}))$ such that

$$\|f^{yy}\|^{U^1_m,\Theta_{m+1}}_{a,a-d_H} \le C(m)\varepsilon_m^{-2/3} \tag{2.38}$$

for $a \in [-d_c,d]$.

The main step of the lemma's proof (and the heart of the Theorem's proof) is an estimate for small denominators which arise in the construction (We gave a sketch of the proof of the estimate above in Part 2.1 during the discussion Step 4 of the proof.)

Remark. In fact, in Part 3 we prove more and show that in the factorization $f^{yy} = f^{0yy} + F^{yy}$ of the operator f^{yy} on the diagonal (with respect to the complex basis $\{(\varphi_j^+ \pm i\varphi_j^-)/\sqrt{2}\}$) part f^{0yy} and the out of the diagonal part F^{yy}, the operator

F^{yy} satisfies the estimate (2.38) with $-d_H$ replaced by $d_1 - 1 - d_H$. This "smoothing property" of the construction can be used to obtain an additional smoothness of the invariant tori we stated in Remark 3, §1 (see [K6]).

Now, with the hamiltonian F in hands, we should check that the transformation S_m and the term $O(\varepsilon_{m+1})$ in (2.30) are smooth enough. To state the result we need some extra notations. We denote by $\Pi_{\mathcal{Y}}, \Pi_\Theta$ the projectors

$$\Pi_{\mathcal{Y}} : \mathcal{Y}^c \times \Theta_0 \longrightarrow \mathcal{Y}^c, \quad \Pi_\Theta : \mathcal{Y}^c \times \Theta_0 \longrightarrow \Theta_0 ,$$

and denote by Π_q, Π_ξ, Π_y the projectors of $\mathcal{Y}^c = (\mathbb{C}^n/2\pi\mathbb{Z}^n) \times \mathbb{C}^n \times Y^c$ on the first, second and third factor respectively.

In Lemma 2.3 and everywhere below we treat the difference of two close $\mathcal{Y}$-valued (or $\mathbb{T}^n$-valued) maps as a E-valued ($\mathbb{R}^n$-valued) map, where $E = \mathbb{R}^{2n} \times Y$.

Lemma 2.3. If $\varepsilon_0 << 1$, then for $\theta \in \Theta_{m+1}$

a) the time-one shift along trajectories of (2.28) defines an analytic map $S_m : O_m^{3c} \to O_m^c$, which is real for real arguments.

b) The map $((q,\xi,y),\theta) \longmapsto S_m$ is Lipschitz-close to the $\mathcal{Y}$-projection, that is

$$|S_m - \Pi_{\mathcal{Y}}|_{E_{d_c}}^{O_m^{3c};\Theta_{m+1}} \le \varepsilon_m^\rho . \tag{2.39}$$

More precisely,

$$|\Pi_q \circ (S_m - \Pi_{\mathcal{Y}})|^{O_m^{3c};\Theta_{m+1}} \le C(m)\varepsilon_m^{1/3} , \tag{2.40}$$

$$|\Pi_\xi \circ (S_m - \Pi_{\mathcal{Y}})|^{O_m^{3c};\Theta_{m+1}} \le C(m)\varepsilon_m , \tag{2.41}$$

$$\|\Pi_y \circ (S_m - \Pi_{\mathcal{Y}})\|_{d_c}^{O_m^{3c};\Theta_{m+1}} \le C(m)\varepsilon_m^{2/3} . \tag{2.42}$$

c) The restriction of S_m on O_{m+1} is a canonical transformation which transforms equations (2.9) – (2.11) on the domain O_m into Hamiltonian equations with a hamiltonian $\mathcal{H}_{m+1}$ of the form (2.2) with $m := m+1$ on the domain O_{m+1}.

The lemma is proved in §5.

Now we are ready to make the last **Step 5** of the proof: to construct the limiting transformation $S_0 \circ S_1 \circ \cdots$ and to prove that this transformation integrates the equations (1.11). Let us set $\Theta = \cap \Theta_m$. Then Θ is a Borel set. By the estimate (2.1),

$$\mathrm{mes}(\Omega_0 \setminus \Theta[I]) \le K^{-1}\gamma\delta_a^n \quad \forall I . \tag{2.43}$$

For $\theta \in \Theta$ and $0 \le r \le N$ let us denote by Σ_N^r the map

$$\Sigma_N^r(\cdot;\theta) = S_r(\cdot;\theta) \circ \cdots \circ S_{N-1}(\cdot;\theta) : O_N^c \longrightarrow O_r^c$$

(as usual, Σ_r^r is the identity mapping).

Lemma 2.4. For all $r, m \ge 0$

$$\|\Sigma_{r+m}^r - \Pi_{\mathcal{Y}}\|_{d_c}^{O_{r+m};\Theta_{m+1}} \le 3\varepsilon_r^\rho , \tag{2.44}$$

where $\mathcal{O}_{r+m} = O^c_{r+m} \cap \mathcal{Y}^c_{d_c}$, endowed with the norm dist_{d_c}.

Proof. Let us denote the l.h.s. in (2.44) by D^r_{r+m}. We can rewrite the identity $\Sigma^r_{r+m}(\mathfrak{h},\theta) = S_r(\Sigma^{r+1}_{r+m}(\mathfrak{h};\theta);\theta)$ in the form

$$\Sigma^r_{r+m} - \Pi_{\mathcal{Y}} = (S_r - \Pi_{\mathcal{Y}}) \circ (\Sigma^{r+1}_{r+m} \times \Pi_\Theta) + (\Sigma^{r+1}_{r+m} - \Pi_{\mathcal{Y}}) .$$

So by (2.39) we get the estimate

$$D^r_{r+m} \le \varepsilon^\rho_r (D^{r+1}_{r+m} + 2) + D^{r+1}_{r+m} .$$

As $D^{r+m}_{r+m} = 0$, then the lemma's assertion results by induction. □

Let us denote by O^c the set

$$O^c = U(\delta_0/2) \times \{0\} \times \{0\} \subset \mathcal{Y}^c .$$

This set contains T^m_0 and is contained in each domain O^c_m because $\delta_m > \frac{1}{2}\delta_0$ for all m.

Lemma 2.5. If $\varepsilon_0 << 1$, then for each fixed m and for $N \to \infty$ the maps $\Sigma^m_{m+N}|_{O^c\times\Theta} : O^c \times \Theta \to \mathcal{Y}^c_{d_c}$ converge to a map Σ^m_∞ such that

a) for each $\theta \in \Theta$ the map $\Sigma^m_\infty(\cdot;\theta) : O^c \to \mathcal{Y}^c_{d_c}$ is complex-analytic;

b) for each $m \le p < \infty$ and all $\theta \in \Theta$

$$\Sigma^m_p(\cdot;\theta) \circ \Sigma^p_\infty(\cdot;\theta) = \Sigma^m_\infty(\cdot;\theta) ; \tag{2.45}$$

c) for large m the map Σ^m_∞ is close to the $\mathcal{Y}$-projection, because

$$\|\Sigma^m_\infty - \Pi_{\mathcal{Y}}\|^{O^c\times\Theta,\text{Lip}}_{d_c} \le 3\varepsilon^\rho_m \quad \forall m , \tag{2.46}$$

moreover, for $\mathfrak{h}$ in O^c and θ in Θ one has

$$\|\Pi_y \circ \Sigma^m_\infty(\mathfrak{h};\theta)\|_{d_c} \le \varepsilon^{1/3+\rho}_m , \quad |\Pi_\xi \circ \Sigma^m_\infty(\mathfrak{h};\theta)| \le \frac{1}{4}\varepsilon^{2/3}_m ; \tag{2.47}$$

d) in particular, $\Sigma^m_\infty(T^m_0) \subset O^m$.

Proof. Let $\mathfrak{h}_0 \in O^c$ and for $j \ge 1$ let $\mathfrak{h}_j = \Sigma^m_{m+j}(\mathfrak{h}_0;\theta)$. Then by (2.39), (2.44)

$$\begin{aligned} \text{dist}_{d_c}(\mathfrak{h}_{N+1},\mathfrak{h}_N) &= \text{dist}_{d_c}\Big(\Sigma^m_{m+N}(S_{m+N}(\mathfrak{h}_0;\theta)),\Sigma^m_{m+N}(\mathfrak{h}_0;\theta);\theta\Big) \\ &\le (1+3\varepsilon^\rho_m)\varepsilon^\rho_{m+N} \le 2\varepsilon^\rho_{m+N} . \end{aligned} \tag{2.48}$$

So the sequence $\{\mathfrak{h}_j\}$ is fundamental and converges to a point $\mathfrak{h}_\infty \in \mathcal{Y}^c_{d_c}$. The r.h.s. of (2.48) does not depend on $\mathfrak{h}_0$. So the sequence $\{\Sigma^m_{m+N}(\cdot;\theta)\}$ converges uniformly in O^c to an analytic map

$$\Sigma^m_\infty(\cdot;\theta) : O^c \longrightarrow \mathcal{Y}^c_{d_c} ,$$

which sends $\mathfrak{h}_0$ to $\mathfrak{h}_\infty$. The relations (2.45) hold and the items a), b) are proved.

The estimate (2.46) results from (2.44) by going to the limit.

To prove (2.47) let us take $\mathfrak{h} \in O^c$, set $\mathfrak{h}^{m+N+1} = \mathfrak{h}$ and for $m \le j \le m+N$ denote $\mathfrak{h}^j = \Sigma^j_{m+N+1}(\mathfrak{h}^{m+N+1};\theta) \in O^c_j$. Then $\mathfrak{h}^j = S_j(\mathfrak{h}^{j+1};\theta)$ and by (2.42)

$$\left\|\Pi_y \mathfrak{h}^j\right\|_{d_c} \le \left\|\Pi_y \mathfrak{h}^{j+1}\right\|_{d_c} + C(m)\varepsilon_j^{2/3} \le \left\|\Pi_y \mathfrak{h}^{j+1}\right\|_{d_c} + \frac{1}{2}\varepsilon_j^{\rho+1/3}\,.$$

As $\Pi_y \mathfrak{h}^{m+N+1} = 0$, then $\|\Pi_y \mathfrak{h}^m\|_{d_c} \le \varepsilon_m^{\rho+1/3}$. So $\left\|\Pi_y \circ \Sigma^m_{m+N+1}(\mathfrak{h};\theta)\right\|_{d_c} \le \varepsilon_m^{\rho+1/3}$, and going to the limit when $N \to \infty$ one gets the first estimate in (2.47). The second one results from (2.41), because by the latter estimate $|\Pi_\xi \mathfrak{h}^j| \le |\Pi_\xi \mathfrak{h}^{j+1}| + C(j)\varepsilon_j$, and $\Pi_\xi \mathfrak{h}^{m+N+1} = 0$ because $\mathfrak{h}^{m+N+1} = \mathfrak{h} \in O^c$.

The statement d) results from (2.47). □

As $\Lambda_0(\omega, I) = \omega$ and $\Lambda_{r+1} = \Lambda_r + \varepsilon_0 h_r^{0\xi}$, where the vector-function $h_r^{0\xi}$ corresponds to the hamiltonian $\mathcal{H}_m$ with $m = r$ (see (2.18)), then

$$\Lambda_r(\omega, I) = \omega + \varepsilon_0 h_0^{0\xi} + \varepsilon_1 h_1^{0\xi} + \cdots + \varepsilon_{r-1} h_{r-1}^{0\xi}\,. \tag{2.49}$$

By the estimate (2.27)

$$\left|\varepsilon_j h_j^{0\xi}\right|^{\Theta,\mathrm{Lip}} \le C(j)\varepsilon_j^{1/3}\,, \tag{2.50}$$

so the maps Λ_r converge to the Lipschitz map Λ_∞, given by the series

$$\Lambda_\infty = \omega + \varepsilon_0 h_0^{0\xi} + \varepsilon_1 h_1^{0\xi} + \cdots\,, \tag{2.51}$$

and

$$|\Lambda_\infty - \omega|^{\Theta,\mathrm{Lip}} \le C\varepsilon_0^{1/3}\,. \tag{2.52}$$

Let us fix $\theta \in \Theta$ and for $m \ge 0$ denote $\omega_m = \Lambda_m(\theta)$. Then

$$|\omega_m - \omega_\infty| \le C(m)\varepsilon_m^{1/3}\,. \tag{2.53}$$

For $q \in \mathrm{T}^n$ (and θ fixed) we consider the curve $[0,1] \ni t \longmapsto \mathfrak{h}_\infty(t) = (q + t\omega_\infty, 0, 0)$ on the torus T_0^n. By the statement d) of the last lemma the map $\Sigma^m_\infty(\cdot;\theta)$ transforms it into a curve $\mathfrak{h}_m(t) = \big(q_m(t), \xi_m(t), y_m(t)\big)$ in the domain O_m. For large m the latter curve is close to the initial one, because

$$\left|\mathrm{dist}\big(q_m(t), q + t\omega_\infty\big)\right| \le 3\varepsilon_m^{\rho} \tag{2.54}$$

by (2.46) and

$$\|y_m(t)\|_{d_c} \le \varepsilon_m^{1/3+\rho}\,, \quad |\xi_m(t)| \le \frac{1}{4}\varepsilon_m^{2/3} \tag{2.55}$$

by (2.47).

We are going to prove that the curve $\mathfrak{h}_0(t)$, constructed as above with $m = 0$, is a solution of the initial equations (1.11) (coinciding with the equations (2.9) – (2.11) for $m = 0$). To do it we start with analysis of a strong solution $\mathfrak{h}(t) = (q, \xi, y)(t)$

$(0 \le t \le 1)$ of the system (2.9) – (2.11) with hamiltonian $\mathcal{H}_m$ in the domain O_m. Taking the inner product in Y_d of equation (2.11) by $y(t)$ and using (1.10) and (2.8), we obtain the estimate

$$\frac{1}{2}\frac{d}{dt}\|y(t)\|_d^2 = \varepsilon_m \langle J\nabla_y H_m, y\rangle_d + \langle J\nabla_y H^3, y\rangle_d$$

$$\le \varepsilon_m \varepsilon_m^{-1/3} C_*(m)\|y\|_d + \|y\|_d K_1\left(\|y\|_d^2 + \|\xi\|\right).$$

Thus

$$\|y(t)\|_d \le \|y(0)\|_d + \frac{1}{3} t\varepsilon_m^{1/3} \quad \text{for} \quad 0 \le t \le 1. \tag{2.56}$$

Similar from equations (2.9), (2.10) we get that

$$|\xi(t)| \le |\xi(0)| + \frac{1}{3} t\varepsilon_m^{2/3}, \quad \operatorname{dist}\left(q(t), q(0) + t\omega_m\right) \le Ct\varepsilon_m^{\rho} \quad (0 \le t \le 1). \tag{2.57}$$

So if

$$\|y(0)\|_d \le \frac{1}{3}\varepsilon_m^{1/3}, \quad |\xi(0)| \le \frac{1}{3}\varepsilon_m^{2/3}, \quad q(0) \in \mathrm{T}^n, \tag{2.58}$$

then the solution $\mathfrak{h}(t)$ stays inside O_m^1 for $0 \le t \le 1$.

If $\mathfrak{h}^m(t)$ is a weak solution of the system (2.9) – (2.11) such that $\mathfrak{h}^m(0) = \mathfrak{h}_m(0)$, then the estimates (2.58) hold by (2.54) – (2.55) with $t = 0$. So by Proposition 4.4 from Part 1 the solution $\mathfrak{h}^m(t)$ exists for $0 \le t \le 1$ and satisfies the estimates (2.56), (2.57). By the inequalities (2.53) – (2.55) this solution is close to the curve $\mathfrak{h}_m$:

$$\operatorname{dist}_d\left(\mathfrak{h}^m(t), \mathfrak{h}_m(t)\right) \le C\varepsilon_m^{\rho} \quad \text{for} \quad 0 \le t \le 1.$$

The mapping $\Sigma_m^0(\cdot;\theta)$ sends the curve $\mathfrak{h}_m$ to $\mathfrak{h}_0$, because $\Sigma_m^0 \circ \Sigma_\infty^m = \Sigma_\infty^0$; as it is canonical, it also transforms the solution $\mathfrak{h}^m$ to the weak solution $\mathfrak{h}^0$ of (2.9) – (2.11) with $m = 0$ and $\mathfrak{h}^0(0) = \mathfrak{h}_0(0)$. This map is Lipschitz by Lemma 2.4. So $\operatorname{dist}_d\left(\mathfrak{h}^0(t), \mathfrak{h}_0(t)\right) \le C'\varepsilon_m^{\rho}$ for $0 \le t \le 1$ and arbitrary m. Hence $\mathfrak{h}^0 = \mathfrak{h}_0$ and $\Sigma_\infty^0(\mathfrak{h}_\infty(z);\theta)$ is a weak in $\mathcal{Y}_d$ solution of the initial Hamiltonian system.

Now the assertions a) – c) of the theorem result by setting

$$\Sigma(q;\theta) = \Sigma_\infty^0(q,0,0;\theta), \quad \omega' = \omega_\infty. \tag{2.59}$$

The estimate (1.16) results from (2.43) and (1.2), the estimate (1.17) and the statement of Remark 5 result from (2.46) and (1.18) follows from (2.52).

To prove assertion d) we use the fact that Lyapunov exponents are invariant under a change of the phase variable. So the exponents of a solution $\mathfrak{h}_0(t)$ of equations (2.9) – (2.11) with $m = 0$ are equal to ones of the solution $\mathfrak{h}_m(t) = \left(\Sigma_m^0\right)^{-1}\mathfrak{h}_0(t)$ of the equations with $m = m$. Let $\delta\mathfrak{h} = (\delta q, \delta\xi, \delta y)(t)$ be a strong solution of the system (2.9) – (2.11), linearised above the solution $\mathfrak{h}_m(t)$:

$$\begin{aligned} \delta\dot{q} &= \left[\nabla_\xi\left(\varepsilon_m H_m + H^3\right)\left(\mathfrak{h}_m(t)\right)\right]_*(\delta\mathfrak{h}), \\ \delta\dot{\xi} &= -\left[\nabla_q\left(\varepsilon_m H_m + H^3\right)\left(\mathfrak{h}_m(t)\right)\right]_*(\delta\mathfrak{h}), \\ \delta\dot{y} &= JA_m\delta y + J\left[\nabla_y\left(\varepsilon_m H_m + H^3\right)\left(\mathfrak{h}_m(t)\right)\right]_*(\delta\mathfrak{h}). \end{aligned} \tag{2.60}$$

Taking the inner product in E_d of these equations with $\delta\mathfrak{h}(t)$ we get the inequality:

$$\frac{d}{dt}\|\mathfrak{h}(t)\|_d \leq \varepsilon_m^{\rho}\|\mathfrak{h}(t)\|_d \,.$$

The same estimate holds after the change $t \longrightarrow -t$. So the modules of the exponents of variational equations do not exceed ε_m^{ρ}. As m is arbitrary, they are equal to zero. □

3. Proof of Lemma 2.2 (solving the homological equations)

In §§3 – 5 we write ε, δ instead of ε_m, δ_m and sometimes omit the argument θ. We denote $\mathbb{Z}_0^s = \mathbb{Z}^s\backslash\{0\}$, $\mathbb{Z}_0 = \mathbb{Z}\backslash\{0\}$; write $a \vee b$ for $\max(a,b)$ and $a \wedge b$ for $\min(a,b)$ and write $\langle s\rangle$ for $1 + |s|$, where s is an n-vector. In the deductions estimates we systematically use the assumption $\varepsilon_0 << 1$.

The assertions of the lemma will be proven for $\Theta_{m+1} = \Theta_m\backslash(\Theta^1 \cup \Theta^2 \cup \Theta^3)$, where Θ^p are Borel sets such that for $p = 1,2,3$ and all I

$$\operatorname{mes}\Theta^p[I] \leq \frac{1}{3}K^{-1}\gamma\delta_a^n(m+1)^{-2} \,. \tag{3.1}$$

3.1. Proof of statement a)

By (2.3) for all I the map

$$\Theta_m[I] \ni \omega \longmapsto \omega' = \Lambda_{m+1}(\omega, I) \tag{3.2}$$

is Lipschitz-close to the identical map:

$$|\Lambda_{m+1} - \omega|^{\Theta_m[I],\mathrm{Lip}} \leq C\varepsilon_0^{\bar{\rho}} \,. \tag{3.3}$$

So (3.2) is a Lipschitz homeomorphism changing the Lebesgue measure by a factor ≤ 2. I.e., for each I and each Borel subset $\Omega \subset \Theta_m[I]$ the following estimates hold:

$$\frac{1}{2}\operatorname{mes}\Omega \leq \operatorname{mes}\Lambda_{m+1}(\Omega \times \{I\}) \leq 2\operatorname{mes}\Omega \tag{3.4}$$

(see Appendix C, Theorem C1).

We define the sets Θ_s^1, $s \in \mathbb{Z}_0^n$, as

$$\Theta_s^1 = \left\{\theta \in \Theta_m \mid |\omega'(\theta)\cdot s| \leq \delta_a\left((m+1)^2|s|^n C\right)^{-1}\right\}$$

and define Θ^1 as the union of all Θ_s^1. Out of Θ^1 we have

$$|\omega'(\theta)\cdot s| \geq \delta_a(m+1)^{-2}|s|^{-n}C^{-1} \quad \forall s \in \mathbb{Z}_0^n \,.$$

By (3.4) the measure of the set $\Theta^1[I]$ can be estimated via the measure of its image under the map (3.1). As the measure of the strip $\{\omega' \mid |\omega' \cdot s| \le \gamma\}$ does not exceeds $C'\gamma\,|s|^{-1}\,\delta_a^{n-1}$, then

$$\begin{aligned}
\operatorname{mes}\Theta^1[I] &\le \sum_{s\in\mathbf{Z}_0^n} \operatorname{mes}\Theta^1_s[I] \\
&\le 2\sum_{s\in\mathbf{Z}_0^n} \operatorname{mes}\{\omega' \mid |\omega'\cdot s| \le \delta_a(m+1)^{-2}\,|s|^{-n}\,C^{-1}\} \\
&\le \frac{\delta_a^n}{C(m+1)^2}\sum_{s\in\mathbf{Z}_0^n} C_1\,|s|^{-n-1} \le \delta_a^n C_2 C^{-1}(m+1)^{-2}\,.
\end{aligned}$$

So condition (3.1) is satisfied if $C >> 1$. For $\theta \in \Theta_m\backslash\Theta^1$ and $q \in U_m^1$ the solutions of equations (2.31) are given by convergent trigonometric series and satisfy the estimates (2.36) (see [A, Sec. 4.2] and Lemmas B1, B2 in Appendix B below).

3.2. Proof of statement c)

We turn to the equation (2.33) (a proof of the assertion b) concerning equation (2.32) is much simpler, a sketch of it is given below).

For $j \in \mathbb{N}$ we set

$$w_j = (\varphi_j^+ + i\varphi_j^-)/\sqrt{2}\,,$$

$$w_{-j} = (\varphi_j^+ - i\varphi_j^-)/\sqrt{2}\,.$$

The system of the vectors $\{w_j\lambda_{|j|}^{(-s)} \mid j \in \mathbf{Z}_0\}$ forms a Hilbert basis of the complex space Y_s^C. The operator JA_{m+1} is diagonal in this basis and

$$JA_{m+1}(\theta)w_j = i\lambda'_j(\theta)w_j \quad \forall j \in \mathbf{Z}_0\,,$$

where $\lambda'_{-j} = -\lambda'_j$ and

$$\lambda'_j(\theta) = \lambda_j(\theta) + \beta_{jm+1}(\theta) \quad \text{for } j \in \mathbb{N}\,.$$

By (2.6)

$$\left|\lambda'_j - (\operatorname{sgn} j)\lambda_{|j|}\right|^{\Theta_m,\mathrm{Lip}} \le C\varepsilon_0^{\beta}\,|j|^{d_H}\,. \tag{3.5}$$

The elements of the matrix $\{f_{kj}(q)\}$ of the operator $f^{yy}(q)$ in the basis $\{w_j\}$ are given by the formula

$$f_{kj} = \langle f^{yy}w_j, w_{-k}\rangle\,,$$

and similar for the matrix $\{h_{jk}\}$ of the operator h^{yy}. So we can apply the quadratic forms, corresponding to the operators in the l.h.s. and r.h.s. of equation (2.33), to vectors w_j, w_{-k} and obtain the equations on the matrix elements $f_{jk}(q)$:

$$\frac{\partial}{\partial\omega'}f_{kj}(q;\theta) + (\lambda'_j(\theta) - \lambda'_k(\theta))f_{kj} = h_{kj}(q;\theta)\,. \tag{3.6}$$

For a vector-function $\eta(q)$ on the torus T^n we denote by $\hat{\eta}(s)$, $s \in Z^n$, its Fourier coefficients: $\eta(q) = \sum \hat{\eta}(s) e^{iq\cdot s}$. By (2.26) and Lemma B1

$$\left\| \hat{h}^{yy}(s) \right\|_{d,d_c}^{\Theta_m, \mathrm{Lip}} \leq C(m) \varepsilon_m^{-2/3} \exp(-\delta |s|) . \tag{3.7}$$

We remind that the mean-values of the diagonal terms $h_{jj}(q)$ of the operator $h^{yy}(q)$ vanish (see the discussion before Lemma 2.1). So

$$\hat{h}_{jj}(0) = 0 \quad \forall j . \tag{3.8}$$

The equations (3.6) are equivalent to the following relations on the Fourier coefficients:

$$i(\omega' \cdot s + \lambda'_j - \lambda'_k) \hat{f}_{kj}(s) = \hat{h}_{kj}(s) , \tag{3.9}$$

where $\omega' = \Lambda_{m+1}(\theta)$. We split the operator f^{yy} on the diagonal part f^{0yy} and the out of diagonal part F^{yy}. So

$$f^{yy} = f^{0yy} + F^{yy}$$

and $f^0_{kj}(q) = 0$ ($F_{kj}(q) = 0$) if $k \neq j$ (if $k = j$). Then by (3.8), (3.9) $f^0_{jj}(0) = 0$ and

$$f^0_{jj}(s) = -i\hat{h}_{jj}(s)/s \cdot \omega' , \quad \text{if} \quad s \neq 0 .$$

For θ out of Θ^1 we have $|s \cdot \omega'| > \delta_a (m+1)^{-2} |s|^{-n} C^{-1}$. So the estimate (2.38) for the diagonal part f^{0yy} results in the same trivial way as above in §3.1.

Now we should obtain the estimate (2.38) for $F^{yy}(q)$. We denote

$$D(k,j,s;\theta) = \begin{cases} i \ , j = k \ , \\ i(\omega' \cdot s + \lambda'_j - \lambda'_k) \ , j \neq k \ . \end{cases}$$

The Fourier coefficients $\hat{F}_{kj}(s)$ of the matrix element $F_{kj}(q)$ ($k \neq j$) of the operator $F^{yy}(q)$ are equal to

$$\hat{F}_{kj}(s) = \hat{h}_{kj}(s;\theta) D^{-1}(k,j,s;\theta) . \tag{3.10}$$

The key observation for what follows is an estimate on the denominators D. For further purposes we give the estimate with an exact control for the dependence of the involved coefficients on the radius δ_a. We suppose that

$$\varepsilon_0^{\tilde{b}} < C\delta_a^b \quad \text{with some} \quad b > 1 . \tag{3.11}$$

(This estimate clearly holds under the assumptions of Theorem 2.1). The following statement is proven below in §4.

Lemma 3.1. There exists a Borel subset $\Theta^2 \subset \Theta_m$ satisfying (3.1) and a constant $c > 0$ such that for $\Theta \in \Theta_m \backslash \Theta^2$ and all $j \neq k \in Z_0$, $s \in Z^n$ the following estimates

$$\left| D^{-1}(k,j,s;\theta) \right| \leq C(m) \delta_a^{-1} \langle s \rangle^c \left| j^{d_1} - k^{d_1} \right|^{-1} , \tag{3.12}$$

$$\left|D^{-1}(k,j,s;\cdot)\right|^{\Theta_m\setminus(\Theta^1\cup\Theta^2),\mathrm{Lip}} \le C(m)\delta_a^{-2}\langle s\rangle^{2c+1}\left|j^{d_1}-k^{d_1}\right|^{-1} \tag{3.13}$$

hold provided that δ_a and ε_0 are small enough and satisfy (3.11). Here $C(m)$ does not depend on δ_a, and with some abuse of notations we write j^{d_1} for $(\operatorname{sgn} j)\,|j|^{d_1}$, where $j \in \mathbb{Z}_0$.

Below for a map $g(k,p) : \mathbb{Z}_0 \times P \to \mathbb{C}$, where P is an abstract set, we denote by $|g(k,p)|_{\ell^r(k)}$ the ℓ^r-norm of the sequence $\{g(k,p) \mid k \in \mathbb{Z}_0\}$ and treat g as a map from P to $\ell^r(\mathbb{Z}_0)$.

We have to estimate the norm of the operator $F^{yy}(q;\theta)$ for $\theta \in \Theta_m\setminus\Theta^2$. By Lemmas B1, B2 of the Appendix B below this is equivalent to estimate the operator norm of the Fourier coefficients $\hat{F}^{yy}(s)$. For this end we shall do the following:

(1) estimate matrix coefficients $\hat{h}_{kj}(s)$ of the operator $\hat{h}^{yy}(s)$;

(2) estimate the coefficients $\hat{F}_{kj}(s)$ via the relation (3.18);

(3) estimate the norm of the operator $\hat{F}^{yy}(s)$ via norms of its matrix-coefficients $\hat{F}_{kj}(s)$.

Step (1) is rather simple. Indeed, the matrix of the operator $\hat{h}^{yy}(s) : Y_d^c \to Y_{d_c}^c$ with respect to the Hilbert basis $\{\lambda_k^{(-d)}w_k \mid k \in \mathbb{Z}_0\}$ in Y_d^c and the basis $\{\lambda_k^{(-d_c)}w_k \mid k \in \mathbb{Z}_0\}$ in $Y_{d_c}^c$ is equal to $\{\lambda_k^{(d_c)}\hat{h}_{kj}\lambda_j^{(-d)}\}$. As the norm of an operator majorizes ℓ^2-norms of the rows of its matrix, then by (1.5) and Lemma B1 for each j we have:

$$\left|\,|k|^{d_c}\,\hat{h}_{kj}(s)\,|j|^{-d}\right|_{\ell^2(k)} \le C\left\|\hat{h}^{yy}(s)\right\|_{d,d_c} \le C(m)\varepsilon^{-2/3}\exp(-\delta\,|s|)\,. \tag{3.14}$$

Step (2) results from Lemma 3.1, because due to it

$$\left|\hat{F}_{kj}(s)\right| \le T\left|\hat{h}_{kj}(s)\right|\left|j^{d_1}-k^{d_1}\right|^{-1}\,, \tag{3.15}$$

where we denote

$$T = C_1(m+1)^2\langle s\rangle^{2c+1}\,.$$

To make Step (3) we should glue the estimates (3.14), (3.15) to obtain an estimate (2.38) for F^{yy}. In fact, we shall do more and estimate the norm of F^{yy} as an operator from Y_a to $Y_{a+d_1-1-d_H}$. We denote

$$d_e = d + d_1 - 1 - d_H\,.$$

The operator $\hat{F}^{yy}(s)$ from the space Y_d^c with the basis $\{\lambda_j^{(-d)}w_j\}$ to the space $Y_{d_e}^c$ with the basis $\{\lambda_j^{(-d_e)}w_j\}$ has the matrix

$$\{\lambda_k^{(d_e)}\hat{F}_{kj}(s)\lambda_j^{(-d)}\}\,. \tag{3.16}$$

Let us denote by $\pi_{k,j}$ the function $1-\delta_{k,j}$. Then (3.15) implies a trivial estimate for the ℓ^1-norm of the column number j of the matrix (3.16):

$$\left|\lambda_k^{(d_e)}\hat{F}_{kj}(s)\lambda_j^{(-d)}\right|_{\ell^1(k)} \le (CT)\left|\,|k|^{d_c}\,\hat{h}_{kj}(s)\,|j|^{-d}\right|_{\ell^2(k)}\left|\pi_{k,j}\,|k|^{d_1-1}\left|j^{d_1}-k^{d_1}\right|^{-1}\right|_{\ell^2(k)} \tag{3.17}$$

The second factor in the r.h.s. of (3.17) is estimated in (3.14) and the third one equals the square root of

$$\Big(\sum_{k=-\infty}^{|j|-1} + \sum_{k=|j|+1}^{\infty}\Big)\frac{|k|^{2(d_1-1)}}{(k^{d_1}-j^{d_1})^2}.$$

After the substitution $k = |j|\,y$ this sum can be estimated by the integral

$$\frac{C_1}{|j|}\Big(\int_{-\infty}^{1-|j|^{-1}} + \int_{1+|j|^{-1}}^{\infty}\Big)\frac{|y|^{2(d_1-1)}}{(y^{d_1}-1)^2}\,dy \le C_2$$

(we remind that $y^{d_1} = \operatorname{sgn} y\,|y|^{d_1}$). By (3.14), (3.17) and the last estimate, ℓ^1-norm of the column number j of the matrix (3.16) does not exceed

$$L_1 = C_1(m)T\varepsilon^{-2/3}e^{-\delta|s|}.$$

For the ℓ^1-norm of the row number k of the matrix (3.16) we have the estimate similar to (3.17):

$$\Big|\lambda_k^{(d_c)}\hat{F}_{kj}(s)\lambda_j^{(-d)}\Big|_{\ell^1(j)} \le CT\Big||k|^{d_c-1}\hat{h}_{kj}(s)\,|j|^{1-d}\Big|_{\ell^2(j)}\left|\frac{\pi_{k,j}\,|k|^{d_1}}{|j|\,|j^{d_1}-k^{d_1}|}\right|_{\ell^2(j)}. \tag{3.18}$$

As the operator $h^{yy}(q)$ belongs to $\mathcal{L}^s(Y_d^c;Y_{d_c}^c)$ (i.e., is symmetric in Y), then by the interpolation theorem (see Corollary A2),

$$\|h^{yy}\|_{1-d+d_H,1-d}^{U_m} \le 2\,\|h^{yy}\|_{d,d_c}^{U_m} \le C\,C_*(m)\varepsilon^{-2/3},$$

and for the conjugate operator $(h^{yy})^*$ one has the dual estimate

$$\|(h^{yy})^*\|_{d-1,d_c-1}^{U_m} \le C\,C_*(m)\varepsilon^{-2/3}.$$

Thus for θ in Θ_m and all s,k

$$\Big||k|^{d_c-1}\hat{h}_{kj}(s)\,|j|^{1-d}\Big|_{\ell^2(j)} \le \Big|(\hat{h}^{yy})^*(s)\Big|_{d-1,d_c-1} \le C(m)\varepsilon^{-2/3}e^{-\delta|s|}, \tag{3.19}$$

and the first factor in the r.h.s. in (3.18) is estimated. For the second one the following estimates hold:

$$\left|\frac{\pi_{k,j}\,|k|^{d_1}}{|j|\,|j^{d_1}-k^{d_1}|}\right|_{\ell^2(j)}^2 \le \frac{C}{|k|}\Big[\int_{-\infty}^{-|k|^{-1}} + \int_{|k|^{-1}}^{1-|k|^{-1}} + \int_{1+|k|^{-1}}^{\infty}\Big]\frac{dy}{y^2(1-y^{d_1})^2} \le C_1.$$

Thus by (3.18) and (3.19), ℓ^1-norm of the row number k is bounded above by the constant

$$L_2 = C_2(m)T\,\varepsilon^{-2/3}\exp-\delta\,|s|.$$

So the matrix (3.16) of the operator $\hat{F}^{yy}(s) : Y^c_d \to Y^c_{d_e}$ has columns and rows bounded in ℓ^1-norm by $\max(L_1, L_2)$. Hence the norm of the operator is bounded by the same constant; for this classical result see [HLP, Chap. 8] or [HS]. For θ in Θ_{m+1} we have got the estimate

$$\left\|\hat{F}^{yy}(s)\right\|_{d,d_e} \leq C(m)T\,\varepsilon^{-2/3}e^{-\delta|s|}\,,$$

whence by Lemma B2

$$\|F^{yy}(q)\|_{d,d_e} \leq C_1(m)\varepsilon^{-2/3}$$

for q in U^1_m. The Lipschitz constant of the solution F^{yy} can be estimated similarly, thus proving the estimate

$$\|F^{yy}\|^{U^1_m,\Theta_{m+1}}_{a,a+d_1-1-d_H} \leq C(m)\varepsilon_m^{-2/3} \tag{3.20}$$

for $a = d$.

The symmetry of the operator F^{yy} results from the one of the Fourier coefficients $\hat{F}^{yy}(s)$ (formula (3.10)). For $q \in \mathrm{T}^n$ the operator $F^{yy}(q)$ is real because the operators $h^{yy}(q)$, $q \in \mathrm{T}^n$, are real. So $F^{yy}(q) \in \mathcal{L}^s(Y^c_d; Y^c_{d_e})$. Now the validity of the estimate (3.20) for all a in the segment $[-d_e, d]$ results from the estimate for $a = d$, from the symmetry of the operator F^{yy} and from the interpolation theorem (see Corollary A2). Thus assertion c) of the lemma and the remark, following the lemma, are proven.

3.3 Proof of statement b)

Denote by $\hat{f}^y(s)$, $\hat{h}^y(s) \in Y$ the Fourier coefficients of vectors $f^y(q)$, $h^y(q)$, and by $\hat{f}_j(s)$, $\hat{h}_j(s)$ the coefficients of decompositions vectors $\hat{f}^y(s)$, $\hat{h}^y(s)$ in the basis $\{w_j\}$. Then by (2.32)

$$\hat{f}_j(s) = D_1^{-1}(j, s; \theta)\hat{h}_j(s)\,, \tag{3.21}$$

where for $j \in \mathbf{Z}_0$ and $s \in \mathbf{Z}^n$

$$D_1(j, s; \theta) = i(s \cdot \omega' - \lambda'_j)\,.$$

By (2.23) and Lemma B1

$$\left\|\hat{h}^y(s)\right\|^{\Theta_m,\mathrm{Lip}}_{d_c} \leq C(m)\varepsilon^{-1/3}\exp(-\delta\,|s|)\,. \tag{3.22}$$

To estimate D_1^{-1} we use an analog of Lemma 3.1:

Lemma 3.2. There exists a Borel subset $\Theta^3 \subset \Theta_m$ satisfying (3.1) such that for $\theta \in \Theta_m \backslash \Theta^3$

$$|D_1^{-1}| \leq \delta_a^{-1}C(m)\langle s\rangle^{n+1}\,,$$

$$|D_1^{-1}|^{\Theta_m\backslash\Theta^3,\mathrm{Lip}} \leq \delta_a^{-2}C(m)\langle s\rangle^{2n+3}\,, \tag{3.23}$$

with some δ_a-independent constant $C(m)$, provided that δ_a and ε_0 are small enough.

We omit a proof of the lemma because it follows the same lines as the proof of Lemma 3.1 and is much simpler.

By equality (3.21) and estimates (3.22), (3.23)

$$\left\|\hat{f}^y(s)\right\|_{d_c}^{\Theta_{m+1},\mathrm{Lip}} \leq C(m)\langle s\rangle^{2n+3}\varepsilon^{-1/3}\exp(-\delta\,|s|)\,.$$

So by Lemma B2

$$\|f^y\|_{d_c}^{U^1_m,\Theta_{m+1}} + \|\nabla_q f^y\|_{d_c}^{U^1_m,\Theta_{m+1}} \leq C(m)\varepsilon^{-1/3}\,.$$

Now the estimate (2.37) results from the equation (2.32), because due to the latter

$$f^y = (JA_{m+1})^{-1}(\partial f^y/\partial\omega' - h^y)\,.$$

3.4. Refinement of the lemma's statements

The exact control for the dependence on the radius δ_a of the denominator D and its Lipschitz constant, given in Lemma 3.1, allows to control dependence of the solutions of the homological equations on δ_a. In terms of the weighted Lipschitz norms defined in (1.20) the result may be stated as follows.

Lemma 3.3. If $\varepsilon_0^\beta < C\delta_a^b$ with some $b > 1$, then the solutions of the homological equations, constructed in Lemma 2.2, meet the estimates:

$$[f^q]^{U^1_m,\Theta_{m+1}} \leq C(m)\delta_a^{-1}[h^q]^{U_m,\Theta_m}\,,$$

$$\dots$$

$$[f^{yy}]_{a,a-d_H}^{U^1_m,\Theta_{m+1}} \leq C(m)\delta_a^{-1}[h^{yy}]_{d,d-d_H}^{U_m,\Theta_m}\,.$$

The constants $C(m)$ are δ_a-independent.

The proof follows from a simple analysis of the formulae of §§3.1 – 3.3 and from the inequality (1.21) for the weighted norms.

4. Proof of Lemma 3.1 (estimation of the small divisors)

We rewrite the estimate (3.12) we should prove, as follows:

$$|D(k,j,s;\theta)| \geq \delta_a\frac{\left|j^{d_1} - k^{d_1}\right|}{C(m+1)^2\langle s\rangle^c}\,, \tag{4.1}$$

where $k \neq j$ and θ lies outside of $\Theta^1 \cup \Theta^2$ (with the set Θ^2 to be constructed). As

$$\mathrm{Lip}\,D^{-1} \leq \mathrm{Lip}\,D\,(\inf|D|)^{-2}$$

and

$$\mathrm{Lip}\,D(k,j,s;\cdot) \leq C\big(\langle s\rangle + (|j| \vee |k|)^{d_1-1}\big)\,,$$

then (4.1) also implies (3.13). As $D(k,j,s) = D(-j,-k,s) = -D(-k,-j,-s)$, then we may suppose that

$$j > 0\,, \quad j \geq |k|\,, \quad j \neq k\,. \tag{4.2}$$

We denote $\Lambda_{jk} = \lambda'_j - \lambda'_k$ and write (4.1) as

$$|\omega' \cdot s + \Lambda_{jk}| \geq \kappa = \delta_a \frac{j^{d_1} - k^{d_1}}{C_*(m+1)^2 \langle s \rangle^{c_1}} \tag{4.3}$$

with some C_*, c_1 to be found.

If $|s| \leq M_1$ and $j \leq j_1$, then (2.3), (3.5) and assumption 2) of the theorem jointly imply (4.3). So in what follows we may suppose that

$$|s| \geq M_1 \quad \text{or} \quad j \geq j_1\,.$$

Lemma 4.1. If j_1 in the assumption 2) of the theorem is large enough (but depends only on the quantities mentioned in Refinement 1), then

$$|\Lambda_{jk}| \geq C_0^{-1} \left| j^{d_1} - k^{d_1} \right| \quad \text{if} \quad j \geq j_1$$

and

$$|\Lambda_{jk}| \geq C_0^{-1} \delta_a \left| j^{d_1} - k^{d_1} \right|$$

for all j, k as in (4.2).

Proof. Denote $\Lambda^0_{jk} = \lambda_j - \lambda_k$. By the estimate (3.5) and the relations $d_H \leq d_1 - 1$, $|j-k| \geq 1$ we find that

$$\left| \Lambda_{jk} - \Lambda^0_{jk} \right| \leq C \varepsilon_0^{\beta} \left(|j|^{d_H} + |k|^{d_H} \right) \leq C_1 \varepsilon_0^{\beta} \, |j-k|^{d_1 - 1}\,.$$

So in view of the assumption (3.11) it is sufficient to prove the inequalities with Λ_{jk} replaced by Λ^0_{jk}.

The first estimate for Λ^0_{jk} results from the assumption 1) of the theorem (one can take, e.g., $C_0^{-1} = \frac{1}{2} K_1^{-1}$), and implies the second estimate because

$$\left| \Lambda^0_{jk} \right| \geq \delta_a K_3$$

for all $|k| \leq j \leq j_1$ by the assumption 2) (with $s = 0$). □

By (1.2) and (2.3) $|\omega'| \leq 2K$. So due to the last lemma, the estimate (4.3) holds trivially if $j \geq j_1$ and $|s| \leq C^{-1} \left| j^{d_1} - k^{d_1} \right|$, where $C = 4C_0 K$. So it is sufficient to check the estimate in two cases:

a) $j \leq j_1$ and $|s| \geq M_1$

b) $j \geq j_1$ and

$$|s| \geq C_0^{-1} (j^{d_1} - k^{d_1}) \tag{4.4}$$

(observe that in both cases $s \neq 0$).

The set Θ^2 will be constructed as the union $\Theta^2 = \Theta^{2,1} \cup \cdots \cup \Theta^{2,4}$, where $\Theta^{2,3} = \Theta^{2,4} = \emptyset$ if $d_1 > 1$ and each set $\Theta^{2,\ell}$ meets the estimate

$$\operatorname{mes} \Theta^{2,\ell}[I] \leq \frac{1}{12} K^{-1} \gamma \delta_a^n (m+1)^{-2} . \tag{4.5}$$

For k, j, s with k, j as in (4.2) we define the set $\Theta(k, j, s)$ as the union of all $\theta \in \Theta_m \backslash \Theta^1$, violating (4.3) with fixed k, j, s.

Lemma 4.2. For each $I \in \mathcal{J}$

$$\operatorname{mes} \Theta(k, j, s)[I] \leq C \delta_a^n \frac{j^{d_1} - k^{d_1}}{C_*(m+1)^2 \langle s \rangle^{c_1+1}} ,$$

provided that assumption a) or b) holds and j_1, M_1 are large enough.

Proof. For a fixed I denote $\Theta(k, j, s)[I] = \tilde{\Omega}$ and consider the map

$$\tilde{\Omega} \longrightarrow \mathbb{R}^n , \quad \omega \longmapsto \omega'(\theta, I) \equiv \Lambda_{m+1} . \tag{4.6}$$

This map is Lipschitz-close to the identical, so it is a Lipschitz homeomorphism which changes the diameters of sets and their Lebesgue measures no more than by the factor two (see Appendix C). So to estimate $\operatorname{mes} \tilde{\Omega}$ is equivalent to estimate its measure in "ω'-representation", i.e., to estimate the measure of the image Ω' of the map (4.6). To make it we express λ_j', λ_k' and Λ_{jk} (with I fixed) as function of ω' and write Ω' as

$$\Omega' = \{\omega' \mid |\omega' \cdot s - \Lambda_{jk}(\omega')| \leq \kappa\} ,$$

where κ is the r.h.s. of (4.3).

Due to (1.2), $\operatorname{diam} \Omega' \leq 2K \delta_a$ (if $\varepsilon_0 << 1$). So by the Fubini theorem to estimate $\operatorname{mes} \Omega'$ it is sufficient to estimate the one-dimensional measure of the intersection of Ω' with every line parallel to some fixed direction. In particular, to the direction given by the vector $S = s\,|s|^{-1}$. The intersection of Ω' with the line $L_\eta = \{\eta + tS \mid t \in \mathbb{R}\}$, $\eta \in \mathbb{R}^n$, is equal to the set

$$\{t \in \mathbb{R} \mid |\Gamma(t)| \leq \kappa\} , \tag{4.7}$$

where

$$\Gamma(t) = \left(\omega' \cdot s + \Lambda_{jk}(\omega')\right) \big|_{\omega' = \eta + tS} .$$

Observe that $(\partial/\partial t)\, \omega' \cdot s = |s|$. So for $t_1 > t_2$

$$\Gamma(t_1) - \Gamma(t_2) \geq |s| (t_1 - t_2) - (t_1 - t_2) \operatorname{Lip} \Lambda_{jk} ,$$

where $\operatorname{Lip} \Lambda_{jk} = \operatorname{Lip}(\omega' \longmapsto \Lambda_{jk})$. By (1.13) and (3.5)

$$\operatorname{Lip} \Lambda_{jk} \leq C \left(j^{d_{1,r}} + \varepsilon_0^{\beta} j^{d_H}\right) \vee \left(|k|^{d_{1,r}} + \varepsilon_0^{\beta} |k|^{d_H}\right) .$$

So if M_1, j_1 are large enough, then

$$\Gamma(t_1) - \Gamma(t_2) \geq \frac{1}{2} |s| (t_1 - t_2) .$$

To prove this statement we start with the case a) and fix M_1 large enough to guarantee the estimate. In the case b) we observe that for $d_1 > 1$ we can find some $\tilde{d} > 0$ such that $d_1 - 1 > \tilde{d} > d_{1,r} \vee d_H$. So

$$\frac{1}{2} |s| \geq C_1^{-1} j^{d_1 - 1} , \quad \operatorname{Lip} \Lambda_{jk} \leq 2C j^{\tilde{d}} ,$$

and the statement is proven. If $d_1 = 1$, then $-d_0 := d_{1,r} \vee d_H < 0$ and we have

$$\frac{1}{2} |s| \geq C_1^{-1}(j - k) , \quad \operatorname{Lip} \Lambda_{jk} \leq 2C |k|^{-d_0} .$$

These estimates readily imply the statement, provided that $j \geq j_1 >> 1$.

Thus, the measure of the set (4.7) is no larger than $2 |s|^{-1} \kappa$.

This estimate jointly with the Fubini theorem implies that $\operatorname{mes} \Omega' \leq (2K\delta_a)^{n-1} 2 |s|^{-1}$. Now the lemma is proven because $\operatorname{mes} \tilde{\Omega} \leq 2 \operatorname{mes} \Omega'$. □

We start with the case a) and consider the set $\Theta^{2,1}$ equal to the union of all $\Theta(k, j, s)$ with k, j, s as in a). By the lemma,

$$\operatorname{mes} \Theta^{2,1} \leq C_1 \delta_a^n C_*^{-1} (m+1)^{-2} \sum \langle s \rangle^{-c_1 - 1} .$$

So the estimate (4.5) holds for $\ell = 1$ if $c_1 > n - 1$ and C_* is large enough.

Now we turn to the case b). We should distinguish two subcases depending on whether $d_1 > 1$ or $d_1 = 1$.

b_1) $d_1 > 1$. Now by (4.4)

$$j \leq C |s|^{d_0} , \quad d_0 = d_1 - 1 > 0 . \tag{4.8}$$

Define $\Theta^{2,2}$ as the union of all the sets $\Theta(k, j, s)$ with k, j, s as in b). By Lemma 4.2,

$$\operatorname{mes} \Theta^{2,2}[I] \leq \frac{C \delta_a^n}{C_* (m+1)^2} \sum_s \langle s \rangle^{-c_1 - 1} \sum_{j,k} (j^{d_1} - k^{d_1}) .$$

By (4.4), (4.8) the inner sum in the r.h.s. may be estimated as follows:

$$\sum_{j,k} (j^{d_1} - k^{d_1}) \leq \sum_{j,k} |s| \leq C \langle s \rangle^{2d_0 + 1} .$$

So the set $\Theta^{2,2}$ satisfies (4.5) if $c_1 > n + 2d_0$ and $C_* >> 1$.

b_2) $d_1 = 1$. If $k \leq 0$, then by (4.4) $j \leq C |s|$. We define $\Theta^{2,2}$ as union of all the sets $\Theta(k, j, s)$ with k, j, s as in b) and $k \leq 0$. For the same reason as above the estimate (4.5) holds for this set.

So it remains to handle the denominators D when

$$0 < k < j\,, \quad j_1 \le j\,. \tag{4.9}$$

Denote $M = j - k$. Then

$$0 < M \le C\,|s| \tag{4.10}$$

by (4.4).

Due to the assumption 1) of the theorem and estimates (3.5), (3.11), (4.10)

$$\left|\Lambda_{jk} - MK_2^0\right| \le C\delta_a\,|s|\,k^{-d_0}\,, \tag{4.11}$$

where $-d_0 = d_{1,r} \vee d_H < 0$. We consider the set of all θ such that

$$\left|\omega'\cdot s - MK_2^0\right| \le \frac{2\delta_a}{C_*(m+1)^2\langle s\rangle^{c_1-1}}$$

and denote by $\Theta^{2,3}$ the union of all these sets with respect to M, s as in (4.10). We can argue as above to prove that

$$\operatorname{mes}\Theta^{2,3}[I] \le \sum_{s\in Z_0^n}\sum_{M=0}^{2C_0|s|}\frac{C\delta_a^n}{C_*(m+1)^2\langle s\rangle^{c_1-2}} \le \frac{C_1\delta_a^n}{C_*(m+1)^2}\sum_{s\in Z_0^n}\langle s\rangle^{-c_1+1}\,.$$

So the estimate (4.5) holds for $\ell = 3$ if $c_1 > n+1$ and $C_* >> 1$.

For θ outside $\Theta^{2,3}$ we have by (4.11) and (4.10):

$$\begin{aligned}|D| &\ge \left|\omega'\cdot s - MK_2^0\right| - \left|\Lambda_{jk} - MK_2^0\right| \ge \frac{2\,\delta_a}{C_*(m+1)^2\langle s\rangle^{c_1-1}} - C\,\delta_a\langle s\rangle k^{-d_0} \\ &\ge \frac{\delta_a(j-k)}{2C_0\,C_*(m+1)^2\langle s\rangle^{c_1}} + C\,\delta_a\langle s\rangle\Big(\frac{1}{C\,C_*(m+1)^2\langle s\rangle^{c_1}} - k^{-d_0}\Big)\,,\end{aligned} \tag{4.12}$$

and the estimate (4.3) holds provided that the second term in the r.h.s. of (4.12) is positive. That is, if $k^{d_0} \ge C\,C_*(m+1)^2\langle s\rangle^{c_1}$.

So it remains to prove (4.1) for m, s as in (4.10) and

$$k^{d_0} < C\,C_*(m+1)^2\langle s\rangle^{c_1}\,. \tag{4.13}$$

To handle this case we define $\tilde\Theta(k,j,s)$ as the set of all θ violating the inequality

$$\left|\omega'\cdot s - \Lambda_{jk}\right| \ge \frac{\delta_a(j-k)}{C_{**}(m+1)^{2(1+d_0^{-1})}\langle s\rangle^{c}} \tag{4.14}$$

(the numbers C_{**} and c will be defined later). Define $\Theta^{2,4}$ as the union of all $\tilde\Theta$ with k,j,s as in (4.9), (4.10), (4.13). Then

$$\operatorname{mes}\Theta^{2,4}[I] \le \frac{C\,\delta_a^n}{C_{**}(m+1)^{2(1+d_0^{-1})}}\sum_s\langle s\rangle^{-c-1}\sum_{j>k>0}(j-k)\,. \tag{4.15}$$

In view of (4.10) the number of different indexes (j,k) satisfying (4.9), (4.13) is majorized by

$$C\,C_*^{1/d_0}(m+1)^{2/d_0}\langle s\rangle^{c_1/d_0}M \le C_1\,C_*^{1/d_0}(m+1)^{2/d_0}\langle s\rangle^{c_1/d_0+1}\,.$$

So the iterated sum in the r.h.s. of (4.15) may be estimated as follows:

$$\sum_s \langle s\rangle^{-c-1}\sum_{j>k>0}(j-k)\le C\sum_s\langle s\rangle^{-c}\sum_{j>k>0}1$$
$$\le C'\sum_s\langle s\rangle^{-c}C_*^{1/d_0}(m+1)^{2/d_0}\langle s\rangle^{c_1/d_0}\langle s\rangle \le CC_*^{1/d_0}(m+1)^{2/d_0}\,,$$

if $c>n+1+c_1/d_0$.

Hence the set $\Theta^{2,4}$ satisfies the estimate (4.5) if C_{**} is large enough (also with respect to the constant C_*).

We define $\Theta^2=\cup_j\Theta^{2,j}$. This set satisfies (3.1) and for θ outside of Θ^2 holds the estimate (4.3) if $d_1>1$, and either the estimate (4.3) or the estimate (4.14) if $d_1=1$. Thus, the lemma is proven.

Remark. Given above proof of Lemma 3.1 is the only part of the theorem's proof, where we use the second assumption of the theorem. The exact control for δ_a-dependence for all the constants appearing in the lemma's proof shows that the numbers j_1, M_1 do not depend on δ_a. So they depend only on the quantities listed in Refinement 1.

5. Proof of Lemma 2.3 (estimation of the change of variables)[12)]

Everywhere in this part we write for short

$$O^j \text{ for } O^{jc}_m\,,\quad U^j \text{ for } U^j_m\,,\quad \Theta \text{ for } \Theta_{m+1}\,.$$

We denote by E_s^{c+} (E_s^{c-}) the space E_s^c endowed with the weighted norm $\|\cdot\|_{(+,s)}$ (respectively $\|\cdot\|_{(-,s)}$), where

$$\|(p,\xi,y)\|^2_{(\pm,s)}=|p|^2+\varepsilon^{\pm4/3}|\xi|^2+\varepsilon^{\pm2/3}\|y\|_s^2\,.$$

The following assertion results from the definition.

Lemma 5.1. For all $s\in\mathbb{R}$ the spaces E_s^{c+} and E_{-s}^{c-} are dual with respect to the bilinear pairing $<\cdot,\cdot>_E$: $E^c\times E^c\to\mathbb{C}$. That is,

$$\|\mathfrak{h}\|_{(+,s)}=\sup_{\|\mathfrak{h}^*\|_{(-,-s)}\le1}\langle\mathfrak{h},\mathfrak{h}^*\rangle_E\,.$$

---

12) The proof of the lemma was simplified and clarified with respect to the original one given in [K1, K2, K8] due to some techniques from [P1]. I am indepted to J. Pöschel for the corresponding discussions.

We denote by $\mathrm{dist}_{(-,s)}$ the metric in $\mathcal{Y}_s^c$ induced by $\|\cdot\|_{(-,s)}$.

Let us write down the system (2.28) in the form

$$\dot{\mathfrak{h}} = \varepsilon\mathcal{F}(\mathfrak{h}) \,, \mathfrak{h} = \mathfrak{h}(t) = (q(t), \xi(t), y(t)) \,, \tag{5.1}$$

where

$$\mathcal{F}(\mathfrak{h}) = J^{\mathcal{Y}}\nabla F(\mathfrak{h}) \,, \quad J^{\mathcal{Y}}(\delta q, \delta\xi, \delta y) = (\delta\xi, -\delta q, J\delta y)$$

(see also Part 1.4). So $\mathcal{F} = (\mathcal{F}^q, \mathcal{F}^\xi, \mathcal{F}^y)$ and $\mathcal{F}^q = \nabla_\xi F$, $\mathcal{F}^\xi = -\nabla_q F$, $\mathcal{F}^y = J\nabla_y F$.

If $\varepsilon_0 << 1$, then for $j = 1, 2, \ldots, 5$ we have

$$\mathrm{dist}_{(-,d)}(O^{j+1}, O^c_m \backslash O^j) \geq C^{-1}(m) \tag{5.2}$$

by the definition of the $\|\cdot\|_{(-,s)}$-norm and of the domains $O^j = O^{jc}_m$. By Lemma 2.2 and Cauchy estimate,

$$\|\varepsilon\mathcal{F}\|^{O^2,\Theta}_{(-,d_c)} \leq C(m)\varepsilon^{1/3} \,. \tag{5.3}$$

So for $\mathfrak{h}(0) = \mathfrak{h} \in O^3$, for $0 \leq t \leq 1$ and $\varepsilon_0 << 1$ the solution $\mathfrak{h}(t)$ of (5.1) satisfies the estimate

$$\|\mathfrak{h}(t) - \mathfrak{h}\|_{(-,d)} \leq C(m)\varepsilon^{1/3} \,. \tag{5.4}$$

Thus the solution of (5.1) depends analytically on $\mathfrak{h} \in O^3$ and stays inside O^2 for $0 \leq t \leq 1$. So statement a) of the lemma is proven.

The estimates (2.40) – (2.42) result from (5.3), (5.4) and imply (2.39). So statement b) is also proven.

It remains to check the last assertion of the lemma.

For $\mathfrak{h} \in O^2$ the following estimate on the tangent map $\mathcal{F}_*(\mathfrak{h})$ results from Lemma 2.2:

$$\|\varepsilon\mathcal{F}_*(\mathfrak{h};\cdot)\|^{\Theta,\mathrm{Lip}}_{(-,-d_c),(-,-d_c)} \leq C(m)\varepsilon^{1/3} \,. \tag{5.5}$$

For $t \in [0,1]$ let us set $\eta(t) = S^t_*(\mathfrak{h})\eta$. Then $\eta(t)$ is the solution of the Cauchy problem

$$\dot{\eta}(t) = \varepsilon\mathcal{F}_*(\mathfrak{h}(t))\eta(t), \quad \eta(0) = \eta \,,$$

where $\mathfrak{h}(t) = S^t(\mathfrak{h})$ (i.e., $\mathfrak{h}(t)$ is the solution of (5.1) with $\mathfrak{h}(0) = \mathfrak{h}$). The curve $\eta(t)$ satisfies the integral equation

$$\eta(t) = \eta + \varepsilon \int_0^t \mathcal{F}_*(\mathfrak{h}(\tau))\eta(\tau)\, d\tau \,.$$

Hence by (5.5) $\|\eta(t) - \eta\|_{(-,-d_c)} \leq \|\eta\|_{(-,-d_c)} C(m)t\varepsilon^{1/3}$ for $0 \leq t \leq 1$. So

$$\left\|S^t_*(\mathfrak{h}) - Id\right\|_{(-,-d_c),(-,-d_c)} \leq tC(m)\varepsilon^{1/3} \,, \tag{5.6}$$

if $0 \leq t \leq 1$.

The functional H_{0m+1} from the decomposition (2.19) of the hamiltonian $\mathcal{H}_m$ is C^1-smooth in O^1 and the hamiltonian F is analytic in O^1. So the Proposition 1.4.3 from Part 1 being applied to the Hamiltonian equation (5.1) implies that

$$\frac{d}{dt}H_{0m+1}\circ S^t = \varepsilon\{F, H_{0m+1}\}\circ S^t .$$

By (2.34) the r.h.s. is equal to $-\varepsilon H_{2m}\circ S^t$. So we can apply Proposition once more to calculate the second derivative:

$$\frac{d^2}{dt^2}H_{0m+1}\circ S^t = -\varepsilon^2\{F, H_{2m}\}\circ S^t .$$

Thus,

$$\begin{aligned} H_{0m+1}\circ S_m = H_{0m+1}\circ S^t \big|_{t=1} &= H_{0m+1} + \frac{d}{dt}H_{0m+1}|_{t=0} \\ &+ \int_0^1 (1-t)\frac{d^2}{dt^2}H_{0m+1}\circ S^t dt \\ &= H_{0m+1} - \varepsilon H_{2m} - \varepsilon^2\int_0^1 (1-t)\{F, H_{2m}\}\circ S^t dt . \end{aligned}$$

Similar

$$\begin{aligned} \big(\varepsilon(H_{2m}+H_{3m}) + H^3\big)\circ S_m &= \varepsilon(H_{2m}+H_{3m}) + H^3 \\ &+ \varepsilon\int_0^1 \{F, \varepsilon(H_{2m}+H_{3m}) + H^3\}\circ S^t dt . \end{aligned}$$

Thus the transformed hamiltonian can be written as

$$\begin{aligned} \mathcal{H}_m\circ S_m = (H_{0m+1} + H^3) + \varepsilon H_{3m} + \varepsilon^2\int_0^1 t\{F, H_{2m}\}\circ S^t dt \\ + \varepsilon^2\int_0^1 \{F, H_{3m}\}\circ S^t dt + \varepsilon\int_0^1 \{F, H^3\}\circ S^t dt . \end{aligned} \tag{5.7}$$

For $j = 1, 2, \dots, 5$ we denote by $\Delta_j H$ the j-th term in the r.h.s. of (5.7) (together with the preceding factor). To prove that the hamiltonian $\mathcal{H}_{m+1} := \mathcal{H}_m \circ S_m$ has the form (2.2) we should check that

$$\Delta_2 H + \dots + \Delta_5 H = \varepsilon_{m+1} H_{m+1} ,$$

where H_{m+1} is a function satisfying estimates (2.7), (2.8) in the domain O^c_{m+1}.

By the assumptions (1.9), (1.10) of the theorem and the estimates of Lemma 2.1, everywhere in O^1 the functionals $H = \varepsilon H^3$ and $H = \varepsilon^2 H_{2m}$, $\varepsilon^2 H_{3m}$ meet the estimates

$$|H|^{O^1,\Theta} \le C(m)\varepsilon^2 , \quad \|\nabla_y H\|^{O^1,\Theta}_{d_c} \le C(m)\varepsilon^{5/3} . \tag{5.8}$$

Lemma 5.2. If an analytic function H satisfies estimates (5.8), then for $\mathfrak{h} \in O^3$ we have

$$|\{F, H\}|^{O^3,\Theta} \le C(m)\varepsilon^{4/3} , \tag{5.9}$$

$$\|\nabla\{F,H\}\|^{O^3,\Theta}_{(+,d_c)} \leq C(m)\varepsilon^{4/3}. \tag{5.10}$$

Proof. By the definition of the Poisson bracket in $\mathcal{Y}$ (see Part 1.4),

$$\{F,H\}(\mathfrak{h}) = \langle J^{\mathcal{Y}}\nabla F(\mathfrak{h}), \nabla H(\mathfrak{h})\rangle_E\,. \tag{5.11}$$

It follows from (5.8) (and the Cauchy estimate) that

$$\|\nabla H\|^{O^2,\Theta}_{(+,d_c)} \leq C(m)\varepsilon^2\,. \tag{5.12}$$

As $J^{\mathcal{Y}}\nabla F = \mathcal{F}$, then the estimates (5.3), (5.12) and Lemma 5.1 jointly imply (5.9).

By (5.11) we have

$$\begin{aligned} d\{F,H\}(\mathfrak{h})\delta\mathfrak{h} &= \langle J^{\mathcal{Y}}(\nabla F(\mathfrak{h}))_*\delta\mathfrak{h}, \nabla H(\mathfrak{h})\rangle_E \\ &\quad + \langle J^{\mathcal{Y}}\nabla F(\mathfrak{h}), (\nabla H(\mathfrak{h}))_*\delta\mathfrak{h}\rangle_E\,. \end{aligned}$$

As the Hessian of a function is a selfadjoint operator, then the r.h.s. is equal to

$$\langle \delta\mathfrak{h}, -(\nabla F)_* J^{\mathcal{Y}}\nabla H + (\nabla H)_* J^{\mathcal{Y}}\nabla F\rangle_E\,.$$

Thus

$$\nabla\{F,H\} = (\nabla H)_* J^{\mathcal{Y}}\nabla F - (\nabla F)_* J^{\mathcal{Y}}\nabla H\,. \tag{5.13}$$

The relations (5.2), (5.12) and Cauchy estimate (see Appendix D with $F = \nabla H$, $O = O^2$ and w in O^3) imply that for $\mathfrak{h} \in O^3$

$$\|(\nabla H)_*(\mathfrak{h})\|_{(-,d),(+,d_c)} \leq C(m)\varepsilon^2.$$

As $\|J^{\mathcal{Y}}\nabla F\|_{(-,d_c)} \leq C(m)\varepsilon^{-2/3}$ by (5.3), then the first term in the r.h.s. of (5.13) is bounded by $C_1(m)\varepsilon^{4/3}$. The second term can be estimated similar. So $\|\nabla\{F,H\}\|_{(+,d_c)} \leq C(m)\varepsilon^{4/3}$ in O^3.

The Lipschitz constant in $\theta \in \Theta$ may be estimated in the same way. So the estimate (5.10) is obtained and the lemma is proven. □

Now we can estimate the Poisson bracket, translated along the flow.

Lemma 5.3. If an analytic function H satisfies (5.8), then for $\mathfrak{h} \in O^5$ and $0 \leq t \leq 1$

$$\left|\{F,H\}\circ S^t\right|^{O^5,\Theta} \leq C_1(m)\varepsilon^{4/3}\,, \tag{5.14}$$

$$\left\|\nabla_y(\{F,H\}\circ S^t)\right\|^{O^5,\Theta}_{d_c} \leq C_1(m)\varepsilon\,. \tag{5.15}$$

Proof. As the analytic map S^t maps O^4 to O^3, then for $\mathfrak{h} \in O^5$

$$\nabla(\{F,H\}\circ S^t) = (S^t)^*(\nabla\{F,H\}\circ S^t)\,. \tag{5.16}$$

By the estimate dual to (5.6) and Lemma 5.1, for $\mathfrak{h} \in O^3$ and $0 \le t \le 1$ we have

$$\|S^t(\mathfrak{h})^*\|_{(+,d_c),(+,d_c)} \le 2\,.$$

So by (5.10) the $E^{c+}_{d_c}$-norm of the r.h.s. in (5.16) does not exceed $C(m)\varepsilon^{4/3}$. Observe that $\nabla_y = \Pi_y \circ \nabla$. As

$$\|\Pi_y\|_{(+,d_c),d_c} \le \varepsilon^{-1/3}\,,$$

then $\|\nabla_y(\{F,H\}) \circ S^t\|_{d_c} \le C(m)\varepsilon$. The Lipschitz constant in θ may be estimated similar, so (5.15) is proven.

The estimate (5.14) results from the observation that the norm and the Lipschitz constant of the map

$$O^4_{(-,d)} \times \Theta \longrightarrow \mathbb{C}\,, \quad (\mathfrak{h},\theta) \longmapsto \{F,H\}$$

both are bounded by $C(m)\varepsilon^{4/3}$ due to (5.2), (5.9) and the Cauchy estimate (the domain O^4 should be given the metric $\operatorname{dist}_{(-,d)}$). □

Thus, we define the functional $\varepsilon_{m+1}H_{m+1}$ as $\Delta_2 H + \cdots + \Delta_5 H$. The terms $\Delta_3 H$, $\Delta_4 H$, $\Delta_5 H$ and their y-gradients can be estimated by Lemma 5.3. For example,

$$\nabla_y \Delta_3 H = \int_0^1 t\nabla_y(\{F, \varepsilon^2 H_{2m}\} \circ S^t)\,dt\,,$$

and by (5.15) $\|\nabla_y \Delta_3 H\|^{O^5,\Theta}_{d_c} \le C_1(m)\varepsilon$. The terms $\Delta_4 H$ and $\Delta_5 H$ meet similar estimates. The term $\Delta_2 H$ was estimated in Lemma 2.1, statement d). So

$$\|\nabla_y \varepsilon_{m+1} H_{m+1}\|^{O^5,\Theta}_{d_c} \le \frac{1}{2} C_*(m+1)\varepsilon^{2/3}_{m+1} + 3C_1(m)\varepsilon\,, \tag{5.17}$$

As $\varepsilon_{m+1} = \varepsilon_m^{1+\rho}$ with $0 < \rho < 1/3$, then the functional H_{m+1} satisfies (2.8), if $\varepsilon_0 << 1$. The estimate (2.7) is fulfilled for similar reasons.

Thus, $\mathcal{H}_{m+1} := \mathcal{H}_m \circ S_m$ has the form (2.2) with $m = m+1$ and Lemma 2.3 is proven.

6. Proof of Refinement 2

Below we write ε for ε_0.

By the definitions of maps Σ^0 and Σ, for $\mathfrak{h} = (q,0,0) \in T^n_0$ we have $\Sigma^0(\mathfrak{h};\theta) = \Pi_{\mathcal{Y}}(q,0,0;\theta) = (q,0,0) \in \mathcal{Y}$ and $\Sigma(\mathfrak{h};\theta) = \Sigma^0_\infty(q,0,0;\theta)$. So we should prove that

$$\|\Sigma^0_\infty - \Pi_{\mathcal{Y}}\|^{T^n_0 \times \Theta, \mathrm{Lip}}_{d_c} \le C\varepsilon\,. \tag{6.1}$$

(We remind that the difference of two close $\mathcal{Y}_{d_c}$-valued maps is treated as an E_{d_c}-valued map.) By (2.45) the map Σ^0_∞ is equal to

$$\Sigma^0_\infty(\mathfrak{h};\theta) = S_0(\cdot;\theta) \circ \cdots \circ S_{m-1}(\cdot;\theta) \circ \Sigma^m_\infty(\mathfrak{h};\theta)\,, \tag{6.2}$$

and by (2.46)

$$\|\Sigma_\infty^m - \Pi y\|_{d_c}^{O^c;\Theta} \leq 3\varepsilon_m^\rho . \tag{6.3}$$

The r.h.s. in (6.3) is smaller than 3ε if m is large enough (i.e., if $m \geq M$, where $M = M(\rho)$ is the positive solution of the equation $(1+\rho)^M \rho = 1$). So to prove (6.1) it is sufficient to check that

$$\|S_j - \Pi y\|_{d_c}^{O^c_{j+1};\Theta_{j+1}} \leq C\varepsilon \quad \forall j \leq M . \tag{6.4}$$

In a similar way,

$$\omega'(\omega, I) = \omega + \varepsilon h_0^{0\xi} + \varepsilon_1 h_1^{0\xi} + \cdots$$

(see (2.49)), and $\left|\varepsilon_j h_j^{0\xi} + \varepsilon_{j+1} h_{j+1}^{0\xi} + \cdots\right| \leq C(j)\varepsilon_j^{1/3}$ (see (2.53)). So to improve the exponent in (1.18), as it is stated in Refinement 2, we should check that

$$\left|\varepsilon_j h_j^{0\xi}\right| \leq C\varepsilon \quad \forall j \leq M . \tag{6.5}$$

To check (6.4), (6.5) we improve the estimates for the solutions of homological equations and for the corresponding transformations S_j, when $j \leq M$. These improvements are based on a simple observation: for small j the functions involved into our constructions have radii analyticity in ξ and y of order one, rather than $\varepsilon_j^{2/3}$ and $\varepsilon_j^{1/3}$.

More explicitly, for the domains Q_m and Q_m^j, where

$$Q_m^j = O(T_0^n, \delta_m^j, \mathcal{Y}_d^c), \quad Q_m = Q_m^0 , \tag{6.6}$$

we shall prove by induction that the hamiltonian $\mathcal{H}_m$ (see (2.2)) may be analytically extended to the domain Q_m and there

$$\mathcal{H}_m = H_{0m}(\mathfrak{h};\theta) + \varepsilon_0 H_{(m)}(\mathfrak{h};\theta) + H^3(\mathfrak{h};\theta) . \tag{6.7}$$

The functions H^3, H_{0m} are the same as in (1.11), (2.7). The function $\varepsilon_0 H_{(m)}$ is an analytic extension of $\varepsilon_m H_m$ and meets the estimate

$$|H_{(m)}|^{Q_m,\Theta_m} \leq C_m, \quad \|\nabla_y H_{(m)}\|_{d_c}^{Q_m,\Theta_m} \leq C_m . \tag{6.8}$$

For $m = 0$ the representation (6.7) coincides with the initial one. Let us suppose that the statement is true for some $0 \leq m \leq m(\rho) - 1$. We denote the terms $\varepsilon_m H_m$, $\varepsilon_m h^q$, $\varepsilon_m h^{1\xi}$ etc. in the decomposition (2.16) by $\varepsilon H_{(m)}$, $\varepsilon h_{(m)}^q$, $\varepsilon h_{(m)}^{1\xi}$ etc. and denote the coefficients $\varepsilon_m f^q$, $\varepsilon_m f^\xi$ etc. of the hamiltonian $\varepsilon_m F$ by $\varepsilon f_{(m)}^q$, $\varepsilon f_{(m)}^\xi$ etc. By repeating the proof of Lemma 2.1 we have for $h_{(m)}^q$, $h_{(m)}^{1\xi}$ etc. the estimates of the items a), b) of Lemma 2.1 with r.h.s. replaced by C_m (now we do not controll the rate of increase in m).

In particular,

$$\varepsilon_m \left|h_m^{0\xi}\right| = \varepsilon \left|h_{(m)}^{0\xi}\right| \le \varepsilon C_m \,. \tag{6.9}$$

For $H^3_{(m)}$ we have an estimate of the form (6.8).

By repeating the proof of Lemma 2.2 we get for $f^q_{(m)}, f^\xi_{(m)}$ etc. the estimates of form (2.36) – (2.38) with the r.h.s. replaced by C^1_m. So after the analytic extension into domain Q^3_m the vector-field of equation (2.28) is no larger than $C^2_m\varepsilon$. So S_m may be (analytically) extended to a map from Q^4_m into Q^3_m and for this extension the estimates of the item a), Lemma 2.4, hold with r.h.s. replaced by $C^3_m\varepsilon$. In particular,

$$\|S_m - \Pi_{\mathcal{Y}}\|_{d_c}^{\mathcal{Q}_{m+1};\Theta_{m+1}} \le C^3_m \varepsilon \,, \tag{6.10}$$

where $\mathcal{Q}_{m+1} = Q_{m+1} \cap \mathcal{Y}^c_{d_c}$.

Hence the transformed hamiltonian $\mathcal{H}_m \circ S_m$ may be extended to the domain Q_{m+1} and has there the form (6.7) with $m := m+1$.

Now the estimates (6.4) and (6.5) result from (6.9), (6.10) with $m = 0, 1, \ldots, M$.

7. On reducibility of variational equations

In the statement of Theorem 1.1 we made no use of the estimates (2.5), (2.6), (2.24) for the quadratic in y part of hamiltonian $\mathcal{H}_m$. These estimates allow one to prove that the variational equations for (1.11) along solutions $z^\varepsilon(t)$ ($\varepsilon = \varepsilon_0$) are reducible to the constant coefficient equations (this reducibility is a typical by-product of the KAM-procedure; see [A1], §5.5.10).

The variational equations for $\delta z = (\delta q, \delta\xi, \delta y) \in E_d$ along the solution $z = z^\varepsilon(t)$ have the form:

$$\begin{aligned}
&\delta\dot q = \varepsilon\big(\nabla_\xi H^0(z)\big)_*\delta z \,, \quad \delta\dot\xi = -\varepsilon\big(\nabla_q H^0(z)\big)_*\delta z \,, \\
&\delta\dot y = J\Big(A(\theta)\delta y + \varepsilon\big(\nabla_y H^0(z)\big)_*\delta z\Big) \,,
\end{aligned} \tag{7.1}$$

where $\varepsilon H^0 = \varepsilon H_0 + H^3$.

Let us denote by $T^n_\varepsilon = T^n_\varepsilon(\omega, I) \equiv \Sigma_{(\omega,I)}(T^n)$ the invariant tori constructed in Theorem 1.1.

Theorem 7.1. Under the assumptions of Theorem 1.1 there exists an analytical mapping $\Phi_1 : T^n_\varepsilon \to \mathcal{L}(E_d, E_d)$ such that the substitution $\delta z = \Phi_1\big(z(t)\big)\delta\mathfrak{h}$, $\delta\mathfrak{h} = (\delta q_0, \delta\xi_0, \delta y_0) \in E_d$, transforms solutions of (7.1) into solutions of equations

$$\delta\dot q_0 = 0 \,, \quad \delta\dot\xi_0 = 0 \,, \quad \delta\dot y_0 = J\bar A_\infty \delta y_0 \,, \tag{7.2}$$

where $\bar A_\infty(\theta)\varphi_j^\pm = \bar\lambda_j(\theta)\varphi_j^\pm$ and $\left|\bar\lambda_j(\theta) - \lambda_j(\theta)\right| \le C\varepsilon j^{d_H}$ for all j.

Proof. The solution $z(t)$ has the form $\Sigma\big(\mathfrak{h}(t);\theta\big)$, where $\theta \in \Theta$ and $\mathfrak{h}(t) = (q + \omega' t, 0, 0) \in T^n_0$. Denote $\mathfrak{h}_m(t) = \Sigma^m_\infty\big(\mathfrak{h}(t)\big)$. The map $(\Sigma^0_m)^{-1}_*\big(\mathfrak{h}_m(t)\big)$ transforms solution of (7.1) into solutions of equation (2.60) (equal to the linearization of equations (2.9) – (2.11) about the solution $\mathfrak{h}_m(t)$).

The map $\Sigma_m^0(\mathfrak{h}_m(t))_*$ is equal to $S_{0*}(\mathfrak{h}_0(t)) \circ S_{1*}(\mathfrak{h}_1(t)) \circ \cdots \circ S_{(m-1)*}(\mathfrak{h}_m(t))$. So by the estimate (5.6) it converge to a map $\Sigma_*(\mathfrak{h}(t)) : E_d \to E_d$ as $m \to \infty$. The map Σ_* transforms solutions of (7.1) into solutions of the limiting equation, which has exactly the form (7.2) with $\tilde{\lambda}_j(\theta) = \lambda_j(\theta) + \beta_{j\infty}(\theta)$. Here $\beta_{j\infty}(\theta) = \lim \beta_{jm}(\theta)$ satisfies the estimate

$$|\beta_{j\infty}|^{\Theta,\mathrm{Lip}} \leq \frac{1}{2} C \varepsilon_0^{\tilde{\rho}} j^{d_H} .$$

By the Refinement 2 (and its proof in Part 5 above) we can choose $\tilde{\rho} = 1$. So the theorem is proven with

$$\Phi_1(z) = \Sigma(\mathfrak{h})_* \big|_{\mathfrak{h}=\Sigma^{-1}(z)} .$$

8. Proof of Theorem 1.2

The proof of the theorem follows the same idea as in the given above in Part 6 proof of Refinement 2: to improve the estimates for the first M changes of the phase variable by a better control for the radii of analyticity of the involved functions. In addition, we modify the first change of phase variable of the recurrent procedure to obtain better estimates for the first transformed hamiltonian $\mathcal{H}_1 = \mathcal{H}_0 \circ S_0$.

We define domains Q_m^j and Q_m ($m \geq 0$, $5 \geq j \geq 1$) as in (6.6) and check that the functions from Step m, where $0 \leq m \leq m_0$ and m_0 is large enough, can be analytically extended to Q_m. Estimates for the norms of the extensions are based on the refined version of Lemma 2.2 ("Solving the homological equations"), given in Lemma 3.4.

For short we write ε for ε_0 and δ for δ_a. We denote by $C, C_1, \ldots$ different ε- and δ-independent constants and use the weighted Lipschitz norms defined in (1.20). To simplify notations, for a function $H(\mathfrak{h};\theta)$ as in (1.9), (1.10) we denote by $[[H]]^{Q,\Theta}$ infinum of the set of all positive K such that

$$[H(\mathfrak{h};\cdot)]^{\Theta,\mathrm{Lip}} \leq K\left(|\xi|^2 + |\xi|\,\|y\|_d + \|y\|_d^3\right), \quad [\nabla_y H(\mathfrak{h};\cdot)]_{d_c}^{\Theta,\mathrm{Lip}} \leq K\left(|\xi| + \|y\|_d^2\right)$$

for all $\mathfrak{h} \in Q$. In particular, $[[H]]^{Q,\Theta} = \infty$ if H contains linear in ξ or quadratic in y terms.

We start with an apriory assumption that the exponent $\tilde{\rho}$ in (2.3), (2.6) may be chosen equal to one,

$$\tilde{\rho} = 1 . \tag{8.1}$$

As in §6, at the first step the functions h^q, h^ξ, h^y, h^{yy} satisfy the estimates (2.21) - (2.24) without ε in a negative degree in the r.h.s.'s:

$$[h^q]^{U_0,\Theta_0}, \ldots, [h^{yy}]_{d,d_c}^{U_0,\Theta_0} \leq C_1$$

(one can take $C_1 = K_1 \delta_0^{-1}$ because Q^c is the δ_0-neighborhood in $\mathcal{Y}_d$ of the torus T_0^n).

As in §2, we construct the transformation S_0 as the time-one shift along trajectories of the Hamiltonian vector-field with a hamiltonian εF, where $F = f^q + \xi \cdot f^\xi + \langle y, f^y \rangle + \langle y, f^{yy} y \rangle$. Then

$$\frac{d^2}{dt^2} H_{01} \circ S^t = \varepsilon^2 \{F, \{F, H_{01}\}\} \circ S^t$$

and

$$\frac{d}{dt}\big(\varepsilon(H_{20} + H_{30}) + \delta H_0^3\big) \circ S^t = \varepsilon \big\{F, \varepsilon(H_{20} + H_{30}) + \delta H_0^3\big\}$$

(see §5). So the transformed hamiltonian $\mathcal{H}_0 \circ S_0 = \mathcal{H}_0 \circ S^t|_{t=1}$ can be written as

$$\begin{aligned}
\mathcal{H}_0 \circ S_0 = {} & (H_{01} + \delta H_0^3 + \varepsilon H_{30}) \\
& + \varepsilon\big(\{F, H_{01}\} + H_{20} + \delta\{F, H_0^3\}\big) \\
& + \varepsilon\delta \int_0^1 \big(\{F, H_0^3\} \circ S^t - \{F, H_0^3\}\big]\, dt \\
& + \varepsilon^2 \int_0^1 (1-t)\{F, \{F, H_{01}\}\} \circ S^t dt \\
& + \varepsilon^2 \int_0^1 \{F, H_{20} + H_{30}\} \circ S^t dt\,.
\end{aligned} \tag{8.2}$$

This decomposition differs from (5.7) because now the term $\varepsilon\delta \int \{F, H_0^3\} \circ S^t dt$ is too large to be neglected (it is of order ε, as we shall see below). So this term, or equivalently its main part equal to $\varepsilon\delta\{F, H_0^3\}$, should be eliminated by a proper choice of the transformation S_0.

For $j = 1, \ldots, 5$ we denote by $\Delta_j H$ the term number j in the r.h.s. of (8.2). We want to construct the hamiltonian F to kill $\Delta_2 H$ "up to negligible terms". By the definition of the Poisson brackets and in view of the assumption (1.22),

$$\begin{aligned}
\{F, H_0^3\} &= -\nabla_q F \cdot \nabla_\xi H_0^3 + \nabla_\xi F \cdot \nabla_q H_0^3 + \langle J\nabla_y F, \nabla_y H_0^3 \rangle \\
&= -\nabla_q f^q \cdot \nabla_\xi H_0^3 + \langle J f^y, \nabla_y H_0^3 \rangle + \mathcal{O}_1\,,
\end{aligned}$$

where $[[\mathcal{O}_1]]^{Q_0^1, \Theta_1} < \infty$. By the assumption (1.22) all the terms in the r.h.s. are $O(\|y\|_d^2 + |\xi|)$. So

$$-\nabla_q f^q \cdot \nabla_\xi H_0^3 + \langle J f^y, \nabla_y H_0^3 \rangle = h^1(q;\theta) \cdot \xi + \langle \mathcal{A}^1(q;\theta) y, y \rangle + \mathcal{O}_2\,, \tag{8.3}$$

where $[[\mathcal{O}_2]]^{Q_0^1, \Theta_1} < \infty$, the vector h^1 depends only on f^q-component of the unknown hamiltonian F and the operator $\mathcal{A}^1$ depends only on f^q and f^{yy}.

As at the Step 2 of the proof (see §2) we write

$$h^1(q;\theta) = h(q;\theta) + h^0(\theta)$$

and

$$\mathcal{A}^1(q;\theta) = \mathcal{A}(q;\theta) + \mathcal{A}^0(\theta)\,,$$

where h^0 is the averaging of h^1 and $\mathcal{A}^0$ is the averaging of the diagonal part (in the complex basis $\{(\varphi_j^+ \pm i\varphi_j^-)/\sqrt{2}\}$) of the operator $\mathcal{A}^1$. So

$$\{F, H_0^3\} = \langle \mathcal{A}y, y\rangle + \langle \mathcal{A}^0 y, y\rangle + h \cdot \xi + h^0 \cdot \xi + \mathcal{O}_1 + \mathcal{O}_2 \,. \tag{8.4}$$

We want to kill the term $\Delta_2 H$ up to the disparity $\mathcal{O}_1 + \mathcal{O}_2 + \langle \mathcal{A}^0 y, y\rangle + h^0 \cdot \xi$. For this end the functions $f^q, \dots, f^{yy}$ should satisfy the homological equations

$$\begin{aligned}
&\partial f^q/\partial\omega' = h^q \,,\\
&\partial f^\xi/\partial\omega' = h^\xi + \delta h \,,\\
&\partial f^y/\partial\omega' - A_1(\theta) J f^y = h^y \,,\\
&\partial f^{yy}/\partial\omega' + f^{yy} J A_1(\theta) - A_1(\theta) J f^{yy} = h^{yy} + \delta\mathcal{A} \,.
\end{aligned}$$

We can solve the first and the third equation just as before. In view of (1.23) and the a priori assumption (8.1), Lemma 3.4 is applicable with $\tilde{\rho} = 1$ and the solutions meet the estimates

$$[f^q]^{U_0^2,\Theta_1}\,,\ [f^y]_{d_c}^{U_0^2,\Theta_1} \le C\delta^{-1} \,. \tag{8.5}$$

By them and (1.22) we have in (8.3)

$$[\mathcal{A}^1]_{d,d_c}^{U_0^3,\Theta_1} \le C_1\delta^{-1}\,,\quad [h^1]^{U_0^3,\Theta_1} \le C_1\delta^{-1} \,. \tag{8.6}$$

(We remind that $\mathcal{A}^1$ and h^1 depend on f^q and f^y only).

By (8.6) and Lemma 3.4 the second and the fourth homological equations also can be solved and

$$[f^{yy}]_{a,a+\Delta d}^{U_0^4,\Theta_1} \le C\delta^{-1}\,,\quad [f^\xi]^{U_0^4,\Theta_1} \le C\delta^{-1} \,. \tag{8.7}$$

In view of (8.5), (8.7) and (1.24) we have in (8.4)

$$[[\mathcal{O}_j]]^{Q_0^5,\Theta_1} \le C\delta^{-1}\,,\quad j = 1, 2 \,. \tag{8.8}$$

Besides, the E_d-norm of the vector-field $\varepsilon\mathcal{F}$ in (5.1) may be estimated by $C\varepsilon\delta^{-1}$ and its time-one shift S^0 maps Q_1 to Q_0^5 in such a way that

$$S^0 = Id + O(\varepsilon\delta^{-1}) \tag{8.9}$$

(more precisely, for $0 \le t \le 1$ the estimates (5.3) – (5.5) hold with the r.h.s.'s replaced by $C\varepsilon\delta^{-1}$ and the usual Lipschitz norms replaced by the weighted ones).

The term $\Delta_2 H$ in the decomposition (8.2) is equal to

$$\begin{aligned}
\Delta_2 H &= \varepsilon\big(\{F, H_{01}\} + H_{20} + \delta\{F, H_0^3\}\big)\\
&= \varepsilon\delta\big(\langle \mathcal{A}^0(\theta) y, y\rangle + h^0(\theta) \cdot \xi + \mathcal{O}_1 + \mathcal{O}_2\big) \,.
\end{aligned} \tag{8.10}$$

The terms $\Delta_3 H$, $\Delta_4 H$ and $\Delta_5 H$ can be readily estimated by (8.5), (8.7) and (8.9):

$$[\Delta_j H]^{Q_1,\Theta_1},\ [\nabla_y \Delta_j H]^{Q_1,\Theta_1}_{d_c} \le C\varepsilon^2\delta^{-1} \le C\varepsilon^{1+\mu},\ j=3,4,5 \tag{8.11}$$

(to get the inequality (8.11) with $j=4$ one should observe that by (8.10) the interior Poisson bracket $\{F, H_{01}\}$ in $\Delta_4 H$ can be estimated without the factor δ^{-1}).

We denote $\Lambda_1' = \Lambda_1 + \varepsilon\delta h^0$, $A_1' = A_1 + 2\varepsilon\delta\mathcal{A}^0$ and $H_{01}' = H_{01} + h^0(\theta)\cdot\xi + \varepsilon\delta\langle\mathcal{A}^0(\theta)y,y\rangle$. Then by (8.8) – (8.11) the transformed hamiltonian $\mathcal{H}_1 = \mathcal{H}_0 \circ S_0$ may be written as

$$\mathcal{H}_1 = (H_{01}' + \varepsilon H_1^3) + \varepsilon_{(1)}H_{(1)}\,,$$

where $\varepsilon_{(1)} = \varepsilon^{1+\mu}$, the operator A_1' satisfies (2.5), (2.6) with $\tilde\rho = 1$, the vector Λ_1' satisfies (2.3) and the functions H_1^3, $H_{(1)}$ meet the estimates

$$[[H_1^3]]^{Q_1,\Theta_1} \le C_1\,,\quad [H_{(1)}]^{Q_1,\Theta_1} + [\nabla_y H_{(1)}]^{Q_1,\Theta_1}_{d_c} \le C_1\,.$$

We construct the transformations S_j, $j\ge 1$, as in §2 and define $\varepsilon_{(j)} = \varepsilon^{1+\mu j}$. For the transformations S_{j-1} and for the transformed hamiltonians $\mathcal{H}_j = \mathcal{H}_{j-1}\circ S_{j-1}$ we shall prove by induction that

$$S_{j-1} = Id + O(\varepsilon_{(j)}\delta^{-1}) \tag{8.12}$$

and

$$\mathcal{H}_j = (H_{0j} + \varepsilon H_j^3) + \varepsilon_{(j)}H_{(j)}\,.$$

Here, as above, $H_{0j} = \xi\cdot\Lambda_j(\theta) + \frac12\langle A_j(\theta)y,y\rangle$,

$$[[H_j^3]]^{Q_j,\Theta_j} \le C_j\,, \tag{8.13}$$

$$[H_{(j)}]^{Q_j,\Theta_j} + [\nabla_y H_{(j)}]^{Q_j,\Theta_j}_{d_c} \le C_j\,, \tag{8.14}$$

and the operator A_j satisfies estimates (2.5), (2.6).

For $j=1$ the estimates (8.12) – (8.14) are proven. We suppose them for $j=j$ and deduce from this assumption that the estimates hold with $j=j+1$.

As in §2 (see Steps 1,2 of the proof) we decompose $\mathcal{H}_j$ as

$$\mathcal{H}_j = H_{0j+1} + \varepsilon H_j^3 + \varepsilon_{(j)}\big(h^q + \xi\cdot h^{1\xi} + \langle y,h^y\rangle + \langle y,h^{yy}y\rangle + H_{3j}\big)\,,$$

where two "integrable terms" have been extracted from $\varepsilon_{(j)}H_{(j)}$ and added to H_{0j}, thus converting H_{0j} to H_{0j+1}. We have the estimates

$$[h^q]^{U_j,\Theta_j},\ldots,[h^{yy}]^{U_j,\Theta_j}_{d,d_c} \le C_j'\,.$$

By Lemma 3.4 solutions of the homological equations (2.31) – (2.33) meet the estimates

$$[f^q]^{U_j^2,\Theta_{j+1}},\ldots,[f^{yy}]^{U_j^2,\Theta_{j+1}}_{a,a+\Delta d} \le C_j\delta^{-1}\,. \tag{8.15}$$

So E_d-norm of the vector-field $\varepsilon_j\mathcal{F}$ in (5.1) may be estimated by $C_j\varepsilon_{(j)}\delta^{-1}$ and the time-one shift S_j along its trajectories maps Q_{j+1} to Q_j in such a way that S_j satisfies (8.12) with $j = j+1$.

The decomposition (5.7) for the transformed hamiltonian takes now the form

$$\begin{aligned}\mathcal{H}_j \circ S_j = &\left(H_{0j+1} + \varepsilon H_j^3 + \varepsilon_{(j)} H_{3j}\right)\\ &+ \varepsilon_{(j)}^2 \int_0^1 t\{F, H_{2j}\} \circ S^t dt + \varepsilon_{(j)}^2 \int_0^1 \{F, H_{3j}\} \circ S^t dt\\ &+ \varepsilon\varepsilon_{(j)} \int_0^1 \{F, H_j^3\} \circ S^t dt\,.\end{aligned}$$

For $\ell = 1, \dots, 4$ we denote by $\Delta_\ell H$ the ℓ-th term in the r.h.s. For $\ell = 2, 3$ we have by (8.14), (8.15)

$$\begin{aligned}&[\Delta_\ell H]^{Q_{j+1},\Theta_{j+1}} + [\nabla_y \Delta_\ell H]_{d_c}^{Q_{j+1},\Theta_{j+1}}\\ &\quad\le \varepsilon_{(j)}^2 C_j \delta^{-1} \le C_j \varepsilon_{(j)}^2 \varepsilon^{-1+\mu} \le C_j' \varepsilon_{(j+1)}\,.\end{aligned}$$

For $\ell = 4$ we have by (8.13)

$$[\Delta_4 H]^{Q_{j+1},\Theta_{j+1}} + [\nabla_y \Delta_4 H]_{d_c}^{Q_{j+1},\Theta_{j+1}} \le C_j \varepsilon\varepsilon_{(j)} \delta^{-1} \le C_j' \varepsilon_{(j+1)}\,.$$

We denote

$$\varepsilon H_{j+1}^3 = \varepsilon H_j^3 + \varepsilon_{(j)} H_{3j}\,, \quad \varepsilon_{(j+1)} H_{(j+1)} = \Delta_2 H + \Delta_3 H + \Delta_4 H\,.$$

The estimates (8.14) with $j = j+1$ have been just proven, and the estimates (8.13), (8.14) jointly imply the estimate (8.13) for H_{j+1}^3.

By our construction, the vector $\Lambda_j(\theta)$ is equal to

$$\Lambda_j(\omega, I) = \omega + \varepsilon\delta h^0(\omega, I) + \sum_{\ell=0}^{j-1} \varepsilon_{(\ell)} \int \nabla_\xi H_{(\ell)}(q, \xi, 0; \omega, I)|_{\xi=0}\, dq/(2\pi)^n\,.$$

So for $j \le m_0$

$$[\Lambda_j(\theta) - \omega]^{\Theta_j, \mathrm{Lip}} \le C_j \varepsilon\,. \tag{8.16}$$

For $m > m_0$ we apply the main scheme (i.e., the one given in §2) without modifications. Now

$$\varepsilon_m \le \varepsilon_{m_0} \le C_{m_0} \varepsilon^{1+\mu m_0}\,.$$

The small denominators D, appearing during solving the homological equations, are of order $\delta^2 > \varepsilon^{2(1-\mu)}$ (see Lemma 3.1). The denominators are large with respect to the magnitude ε_m of the perturbation, and allow one to carry out the proof given in §2. In fact, simple analysis of the proof shows that the factor δ^{-2} in the estimate (3.13) for the inverse denominator D^{-1} appears as the factor δ^{-2} in the r.h.s.'s of the estimates (2.40) – (2.42) of Lemma 2.3 (i.e., the constants $C(m)$ are equal to $\delta^{-2} C^0(m)$, where $C^0(m)$ are ε- and δ_a-independent). Finally, the factor

δ^{-2} appears in the coefficient $C_1(m)$ at the estimate (5.17) for the transformed hamiltonian. Thus, the inequality (5.17)

$$\|\nabla_y \varepsilon_{m+1} H_{m+1}\|_{d_c}^{O_{m+1},\Theta_{m+1}} \le C_*(m+1)\varepsilon_{m+1}^{2/3}$$

holds if $\delta^{-2}\varepsilon_m < \varepsilon_{m+1}^{2/3} = \varepsilon_m^{2(\rho+1)/3}$, or

$$\delta > \varepsilon_m^{\frac{1}{3}(\frac{1}{2}-\rho)} .$$

As $\rho < 1/3$ and $\delta > \varepsilon^{1-\mu}$, then this inequality holds provided that $m > m_* = 17/\mu - 18$ and $\varepsilon = \varepsilon_0$ is small enough. We fix any $m_0 > m_*$ such that

$$\varepsilon_{m_0}^{\rho} < \varepsilon_0 .$$

Then by (2.50), (2.51) and (8.16) the vector ω' is $C\varepsilon$-close to ω (thus the apriory assumption (8.1) is fulfilled); by (6.2) and (8.12) the integrating transformation Σ is $C\varepsilon\delta^{-1}$-close to the embedding Σ^0. So the theorem is proven.

Appendix A. Interpolation theorem

Let X_1 be a real Hilbert space with a Hilbert basis $\{\eta_j | j \in Z_0\}$ (i.e. $\langle \eta_j, \eta_k \rangle_{X_1} = \delta_{j,k}\ \forall j, k$). Let X_2 be a dense Hilbert subspace of X_1 with the Hilbert basis $\{\chi_j^{-1}\eta_j\}$, where $\chi_j \geq C\ \forall j$. Then for $0 \leq \tau \leq 1$ the interpolation space $[X_2, X_1]_\tau$ is a Hilbert space with the Hilbert basis $\{\chi_j^{-1+\tau}\eta_j | j \in Z_0\}$ (so $[X_2, X_2]_0 = X_2$, $[X_2, X_1]_1 = X_1$). In particular if $X_1 = Y_a$, $X_2 = Y_b$, where $b > a$ and Y_a, Y_b are spaces from the scale $\{Y_s\}$ as in §1, then by the conditions (1.5)

$$[X_2, X_1]_\tau = [Y_b, Y_a]_\tau = Y_{\tau a + (1-\tau) b}$$

(one has to take $\eta_j = \varphi_j^+$ for $j > 0$ and $\eta_j = \varphi_{-j}^-$ for $j < 0$). The norms in the spaces are equivalent:

$$K^{-1} \|y\|_{\tau a + (1-\tau) b} \leq |y|_{[Y_b, Y_a]_\tau} \leq K \|y\|_{\tau a + (1-\tau) b} .$$

Such a scale of Hilbert spaces is called an interpolational scale.

For complexifications X_1^c and X_2^c of the spaces X_1, X_2 we set by definition

$$[X_2^c, X_1^c]_\tau = [X_2, X_1]_\tau \otimes_{\mathbb{R}} \mathbb{C}$$

(i.e. interpolation of complexifications is equal to complexification of interpolation). So $[Y_b^c, Y_a^c]_\tau = Y_{\tau a + (1-\tau) b}^c$.

The $\|\cdot\|_s$-norms of a vector $x \in X_\infty$ are connected by the well-known interpolational inequality

$$\|x\|_{\tau a + (1-\tau) b} \leq \|x\|_a^\tau \|x\|_b^{1-\tau} , \quad 0 \leq \tau \leq 1 \tag{A1}$$

(in other words, the function $t \longmapsto \log \|x\|_\tau$ is convex).

The norms of an operator acting in a Hilbert scale are connected in a similar way:

Theorem A1 (interpolation theorem). Let a linear operator $L : Y_\infty^c \to Y_{-\infty}^c$ may be extended to continuous maps $Y_{s_0}^c \to Y_{\ell_0}^c$ and $Y_{s_1}^c \to Y_{\ell_1}^c$. Then $\forall \tau \in [0,1]$ this operator may be extended to the continuous map $Y_{s_\tau}^c \to Y_{\ell_\tau}^c$, where $s_\tau = \tau s_0 + (1-\tau) s_1$, $\ell_\tau = \tau \ell_0 + (1-\tau)\ell_1$ and

$$\|L\|_{s_\tau, \ell_\tau} \leq C \|L\|_{s_0, \ell_0}^\tau \|L\|_{s_1, \ell_1}^{1-\tau}$$

For a proof see [LM, RS].

Corollary A2. Let a linear continuous operator $Y_s^c \to Y_\ell^c$ be symmetric with respect to the pairing $< \cdot, \cdot >$ (i.e. $L \in \mathcal{L}^s(Y_s^c, Y_\ell^c)$). Then $\forall \tau \in [0,1]$ $L \in \mathcal{L}^s(Y_{s_\tau}^c, Y_{\ell_\tau}^c)$ with $s_\tau = \tau(s+1) - \ell$, $\ell_\tau = \tau(s+1) - \ell$, and $\|L\|_{s_\tau, \ell_\tau} \leq C \|L\|_{s,\ell}$.

Proof. We have the equalities: $\|L\|_{-\ell,-s} = \|L^*\|_{-\ell,-s} = \|L\|_{s,\ell}$. Here L^* is the operator, conjugate to L with respect to the pairing $< \cdot, \cdot >$. Now the assertion results from Theorem A1 with $s_0 = s$, $s_1 = -\ell$, $\ell_0 = \ell$, $\ell_1 = -s$. □

Appendix B. Some estimates for Fourier series

Let B be a Banach space with a norm $\|\cdot\|$, B^c be the complexification of B, $M = \{\mu\}$ be a metric space, $\xi > 0$ and

$$G \in \mathcal{A}_M^R(U(\xi); B^c), \quad \|G\|^{U(\xi),M} \leq 1 . \tag{B1}$$

Let us write the Fourier series for G,

$$G(q;\mu) = \sum_{s \in Z^n} \hat{G}(s;\mu) e^{is\cdot q} . \tag{B2}$$

Lemma B1. For every $s \in Z^n$

$$\left\| \hat{G}(s;\cdot) \right\|^{M,\mathrm{Lip}} \leq e^{-\xi|s|} ; \tag{B3}$$

for all s and μ

$$\hat{G}(s,\mu) = \bar{\hat{G}}(-s,\mu) . \tag{B4}$$

An "almost inverse" statement is true:

Lemma B2. If (B3), (B4) hold for all $s \in Z^n$ and $0 < \Delta < \xi$, then the series (B2) converges for $q \in U(\xi - \Delta)$, the map G is analytic and

$$G \in \mathcal{A}_M^R(U(\xi - \Delta); B^c) , \quad \|G\|^{U(\xi-\Delta),M} \leq 4^n \Delta^{-n} .$$

Lemma B3. If (B1) takes place, $0 < 2\Delta < \xi < 1$ and

$$R_{M_*} G(q) = \sum_{|s| \geq M_*} \hat{G}(s;\mu) e^{is\cdot q} ,$$

then

$$\|R_{M_*} G\|^{U(\xi-2\Delta),M} \leq C(n) \Delta^{-n-1} e^{-3M_*\Delta/4} .$$

The proves of the lemmas given in [A, §4.2] for $B = \mathbb{R}^n$, are valid for arbitrary Banach space B.

Appendix C. Lipschitz homeomorphisms of Borel sets

Let $\Omega \subset \mathbb{R}^n$ be a bounded Borel subset and $\Lambda : \Omega \to \mathbb{R}^n$ be a Lipschitz map of a form $\Lambda(a) = a + \Lambda_1(a)$,

$$\mathrm{Lip}\, \Lambda_1 \leq \mu < 1 . \tag{C1}$$

So

$$\mathrm{Lip}\, \Lambda \leq 1 + \mu . \tag{C2}$$

Theorem C. If (C1) takes place than the inverse map Λ^{-1} is well-defined and

$$\operatorname{Lip} \Lambda^{-1} \leq (1-\mu)^{-1} . \tag{C3}$$

For arbitrary Borel subset $\Omega' \subset \Omega$

$$(1-\mu)^n \operatorname{mes} \Omega' \leq \operatorname{mes} \Lambda(\Omega') \leq (1+\mu)^n \operatorname{mes} \Omega . \tag{C4}$$

Proof. The first statement is evident. Indeed, if $\Lambda(x_j) = y_j$, $j = 1, 2$, then $(x_1 - x_2) + (\Lambda_1 x_1 - \Lambda_1 x_2) = y_1 - y_2$ and by (C1) $|x_1 - x_2|^2 \leq \mu |x_1 - x_2|^2 + |x_1 - x_2| |y_1 - y_2|$. So $|x_1 - x_2| \leq (1-\mu)^{-1} |y_1 - y_2|$ and (C3) is proved.

To prove (C4) we extend Λ to a Lipschitz map $\Lambda^c : \mathbb{R}^n \to \mathbb{R}^n$ with the same Lipschitz constant (Kirszbraun's theorem, see [Fe]). Let $\operatorname{mes} \Omega' = a$. Then the upper measure of Ω' is also equal to a. So $\forall \varepsilon > 0$ the set Ω' may be covered by a countable set of balls $B_j \subset \mathbb{R}^n$ in such a way, that the radius of B_j is equal to r_j, and

$$V_1 \sum_{j=1}^{\infty} r_j^n \leq (1+\varepsilon) a$$

(V_1 is the measure of the 1-ball in $\mathbb{R}^n$). As $\operatorname{Lip} \Lambda^c = \operatorname{Lip} \Lambda \leq (1+\mu)$, then $\Lambda(B_j)$ is contained in a ball of radius $(1+\mu) r_j$. As $\Lambda(\Omega') \subset \cup \Lambda(B_j)$, then

$$\operatorname{mes} \Lambda(\Omega') \leq V_1 \Sigma (1+\mu)^n r_j^n \leq (1+\mu)^n (1+\varepsilon) \operatorname{mes} \Omega' .$$

The second inequality in (C4) is proved because $\varepsilon > 0$ may be chosen arbitrarily small.

To prove the first inequality we should consider the map Λ^{-1} and use (C3).

Appendix D. Cauchy estimate

During the proof we systematically use Cauchy estimate in the following form:

Theorem D. Let X_1^c, X_2^c be complex Hilbert spaces, O be a domain in X_1^c and $F : O \to X_2^c$ be a Fréchet-analytic map. Then

$$\|F_*(w)\|_{X_1, X_2} \leq \delta^{-1} \sup \|F\|_{X_2}$$

for each point w lying in O with its δ-neighborhood.

To prove the estimate we fix unit vectors $x \in X_1^c$, $y \in X_2^c$ and consider the map

$$f : \mathbb{C} \ni z \longmapsto \langle F(w + zx), y \rangle_{X_2} .$$

This map is analytic in δ-disc in the complex plain. So by the usual Cauchy estimate

$$\delta^{-1} \sup \|F\|_{X_2} \geq \delta^{-1} \sup |f| \geq |f'(0)| = \langle F_*(w) x, y \rangle_{X_2} ,$$

and the theorem's assertion is proven. □

A natural reformulation of this statement holds for analytic maps, defined in a subdomain of the toroidal space $\mathcal{Y}^c = (\mathbb{C}^n / 2\pi \mathbb{Z}^n) \times \mathbb{C}^n \times Y^c$. We do not formulate the corresponding result, but use it the proof.

List of notations

1. Constants

m – the number of the iteration;

$C, C_1, C_2, \ldots$ – positive constants which arrive in estimates. They are independent on ε and m and are different in different parts of the text.

$K, K_1, \ldots$ – constants which characterize initial data in theorems;

$C(m), C_1(m), \ldots$ – functions of m of the form $C_1 m^{C_2}$;

$C_{*j}, C_{*j}(m)$ – fixed constants and fixed functions of the form $C(m)$;

$e(m) = \frac{1^{-2}+2^{-2}+\cdots+m^{-2}}{2(1^{-2}+2^{-2}+\cdots)}$ (so $e(m) < \frac{1}{2}$ for all m);

$\varepsilon_m = \varepsilon_0^{(1+\rho)^m}$, where $0 < \rho < 1/3$ is fixed;

$\delta_m = \delta_0(1 - e_m)$ (so $\delta_m > \frac{1}{2}\delta_0$ for all m);

$\delta_m^j = (1 - \frac{j}{6})\delta_m + \frac{j}{6}\delta_{m+1}$, $0 \le j \le 5$;

$a \vee b$ $(a \wedge b)$ – maximum (minimum) of real numbers a, b;

$\langle s \rangle = 1 + |s|$ for $s \in \mathbb{Z}^n$.

2. Linear spaces and maps

Y, Z – Hilbert spaces with norms $\|\cdot\|_Y$, $\|\cdot\|_Z$ and inner products $<\cdot,\cdot>_Y$, $<\cdot,\cdot>_Z$;

$\{Y_s | s \in \mathbb{R}\}$ – a scale of Hilbert spaces Y_s, $|\cdot|_{Y_s} = \|\cdot\|_s$, $Y_0 = Y$, $Y_{s_1} \subset Y_{s_2}$ for $s_1 \ge s_2$, Y_s and Y_{-s} are conjugate with respect to the pairing $<\cdot,\cdot> = <\cdot,\cdot>_Y$;

$\{\lambda_j^{(-s)} \varphi_j^{\pm} | j \in \mathbb{N}\}$ – a Hilbert basis of Y_s, $\lambda_j^{(-s)} = (\lambda_j^{(s)})^{-1} > 0$ $\forall j, \forall s$;

Y^c, Y_s^c – the complexifications of the spaces Y, Y_s, the scalar product $<\cdot,\cdot>$ in Y is extended to the complex-bilinear pairing $Y_s^c \times Y_{-s}^c \to \mathbb{C}$, $s \in \mathbb{R}$;

$E_s = \mathbb{R}^n \times \mathbb{R}^n \times Y_s$, E_s^c – the complexification of E_s;

$\mathcal{L}(Y_s^c, Y_\ell^c)$ – the space of linear continuous operators from Y_s^c to Y_ℓ^c provided with the operator norm $\|\cdot\|_{s,\ell}$;

$\mathcal{L}^s(Y_s^c; Y_\ell^c)$ – operators from $\mathcal{L}(Y_s^c; Y_\ell^c)$, symmetric with respect to $<\cdot,\cdot>$.

3. Sets and domains

$\mathbb{Z}_0^s = \mathbb{Z}^s \backslash \{0\}$, $\mathbb{Z}_0 = \mathbb{Z} \backslash \{0\}$, $\mathbb{R}_+ = \{x \in \mathbb{R} | x \ge 0\}$;

$O(Q, \delta, M)$ – δ-neighborhood of a subset Q of a metric space M;

$O(\delta, Z) = O(0, \delta, Z)$ for a Banach space Z;

$\mathfrak{A} \subset \mathbb{R}^n$ – set of parameters a;

Ω_0 – set of frequency vectors $(\omega_1, \ldots, \omega_n)$, equal to some n-dimensional Borel set of "effective radius δ_a" (see (3.1.2));

$\mathcal{J}$ – set of actions $(I_1, \ldots, I_n)$, in Part 3 $\mathcal{J}$ is an abstract metric space, containing additional (with respect to a or ω) parameters of the equation;

$\Theta_j = \{\theta = (\omega, I)\}$, $j = 1, 2, \ldots$ – Borel subsets of $\Theta_0 = \Omega_0 \times \mathcal{J}$, $\Theta_0 \supset \Theta_1 \supset \cdots$;

$\Theta[I] = \{\omega \in \Omega | (\omega, I) \in \Theta\}$ for $\Theta \subset \Omega \times \mathcal{J}$ and arbitrary $I \in \mathcal{J}$,

$\mathcal{Y}_s = \mathbb{T}^n \times \mathbb{R}^n \times Y_s$, $\mathcal{Y} = \mathcal{Y}_0$, tangent space to $\mathcal{Y}_s$ at a point $\mathfrak{h} \in \mathcal{Y}_s$ is identified with

$E_s = \mathbb{R}^n \times \mathbb{R}^n \times Y_s$;
$\mathcal{Y}_s^c = (\mathbb{C}^n/2\pi\mathbb{Z}^n) \times \mathbb{C}^n \times Y_s^c$;
$U(\delta) = \{\xi \in \mathbb{C}^n/2\pi\mathbb{Z}^n \mid |\operatorname{Im}\xi| < \delta\}$, $U_m = U(\delta_m)$;
$O_m^c = U(\delta_m) \times O(\varepsilon_m^{2/3}, \mathbb{C}^n) \times O(\varepsilon_m^{1/3}, Y_d^c)$;
$O_m^{jc} = U(\delta_m^j) \times O\big((2^{-j}\varepsilon_m)^{2/3}, \mathbb{C}^n\big) \times O\big((2^{-j}\varepsilon_m)^{1/3}, Y_d^c\big)$, $U_m^j = U(\delta_m^j)$;
$O_m = O_m^c \cap \mathcal{Y}_d$;

4. Maps and functions

For a map $G : Q_1 \to Q_2$ (Q_j is a metric space with a distance dist_j, $j = 1, 2$)
$\operatorname{Lip} G = \sup_{x_1 \neq x_2} \frac{\mathrm{dist}_2\left(G(x_1), G(x_2)\right)}{\mathrm{dist}_1(x_1, x_2)}$;
$|G|_{Q_2}^{Q_1,\mathrm{Lip}} = \max\{\sup_{q \in Q_1} |G(q)|_{Q_2}, \operatorname{Lip} G\}$ **if $G : Q_1 \to Q_2$ and Q_2 is a Banach space;**
$\mathcal{A}^R(O_1^c; O_2^c)$ **is the set of Fréchet complex-analytic maps from $O_1^c \subset B_1^c$ to $O_2^c \subset B_2^c$, which map $O_1^c \cap B_1$ into B_2;**
$\mathcal{A}_M^R(O_1^c; O_2^c)$ **is the set of mappings $G : O_1^c \times M \to O_2^c$ such that**
$G(\cdot; m) \in \mathcal{A}^R(O_1^c; O_2^c)$ $\forall m \in M$ **and**

$$|G|_{B_2}^{O_1^c;M} = \sup_{b \in O_1^c} |G(b; \cdot)|_{B_2}^{M,\mathrm{Lip}} < \infty\,;$$

for an antiselfadjoint operator J we denote $\bar{J} = -(J^{-1})$.

Bibliography

[A] Arnold, V.I.: Mathematical methods in classical mechanics. Moscow: Nauka, 1974; English transl. Springer, 1978.

[AA] Arnold, V.I., Avez, A.: Ergodic problems of classical mechanics. Addison–Wesley Publishing Co., 1989.

[A1] Arnold, V.I.: Geometric methods in the theory of ordinary differential equation. Moscow: Nauka, 1978; English transl. Springer, 1983.

[A2] Arnold, V.I.: Proof of a theorem of A.N. Kolmogorov on the conservation of quasiperiodic motions under a small change of the Hamiltonian function. Uspekhi Mat. Nauk 18:5, 13–40 (1963); Russ. Math. Surv. 18:5, 9–36 (1963).

[A3] Arnold, V.I.: Small denominators and problems of stability of motions in classical and celestial mechanics. Uspekhi Mat. Nauk 18:6, 91–192 (1963); Russ. Math. Surv. 18:6, 85–191 (1963).

[AKN] Arnold, V.I., Kozlov, V.V., Neistadt, A.I.: Mathematical aspects of classical and celestial mechanics. Encycl. of mathem. scien., vol. 3. Springer, 1989.

[AlFS] Albanese, C., Fröhlich, J., Spencer, T.: Periodic solutions of some infinite-dimensional hamiltonians associated with non-linear partial differential equations, I and II. Comm. Math. Phys. 116, 475-502 (1988) and 119, 677–699 (1988).

[BFG] Benettin, G., Fröhlich, J., Giorgilli, A.: A Nekhoroshev-type theorems for Hamiltonian systems with infinitely many degrees of freedom. Commun. Math. Phys. 119, 95–108 (1989).

[BGG] Benettin, G., Galgani, L., Giorgilli, A.: A proof of Nekhoroshev's theorem for the stability times in nearly integrable Hamiltonian systems. Celestial Mechanics 37, 1–25 (1985).

[BiK] Bikbaev, R.F., Kuksin S.B.: Periodic boundary-value problem for Sine–Gordon equation, its small Hamiltonian perturbations and KAM-deformations of the finite-gap tori. Algebra i Analiz 4:3, 1–31 (1992). (Russian)

[BoK] Bobenko, A.I., Kuksin, S.B.: Small-amplitude solutions of nonlinear Klein–Gordon equation via Sine–Gordon equation. To appear.

[Bour] Bourgain, J.: Fourier transform restriction phenomena for certain lattice subsets and applications to nonlinear evolution equations. Preprint IHES/M/92/62, Bures-sur-Yvette (1992).

[Bre] Brezis, H.: Operateurs maximaux monotones et semigroupes de contractions dans les espaces de Hilbert. North–Holland, 1973.

[Bre1] Brezis, H.: Periodic solutions of nonlinear vibrating string and duality principles. Bull. Amer. Math. Soc. 8, 409–426 (1983).

[Bru1] Bruno, A.D.: The normal form of a Hamiltonian system. Uspekhi Mat. Nauk 43:1, 23–56 (1988); Russ. Math. Surv. 43:1, 25–66 (1988).

[Bru2] Bruno, A.D.: Normalization of a Hamiltonian system near a cycle or torus. Uspekhi Mat. Nauk 44:2, 49–77 (1989); Russ. Math. Surv. 44:2, 53–89 (1989).

[BS] Berezin, F.S., Shubin, M.A.: The Schrödinger equation, Kluver academic publishers, Dordrecht/Boston/London, 1991.

[Ch-B] Choquet–Bruhat, Y., De Witt-Morette, C.: Analysis, manifolds and physics. North Holland, 1982.

[ChM] Chernoff, P.R., Marsden, J.E.: Properties of infinite-dimensional Hamiltonian systems. Lecture Notes in Math. 425 (1974).

[ChP] Chierchia, L., Perfetti, P.: Second order Hamiltonian equations on T^∞ and almost-periodic solutions. Preprint (1992).

[CW] Craig, W., Wayne, C.E.: Newton's method and periodic solutions of nonlinear wave equation. Preprint (1992).

[DEGM] Dodd, R.K., Eilbeck, J.C., Gibbon, J.D., Morris, H.C.: Solitions and nonlinear wave equations. Academic Press, 1984.

[DPRV] Dodson, M.M., Pöschel, J., Rynne, B.P., Vickers, J.A.G.: The Hausdorf dimension of small divisors for lower dimensional KAM-tori. Proc. R. Soc. Lond. A 439, 359–371 (1992).

[EFM] Ercolani, N., Forest, M.G., McLaughlin, D.: Geometry of the modulational instability. Part 3, Physica 43 D, 349–384 (1990).

[El] Eliasson, L.H.: Perturbations of stable invariant tori. Ann. Sc. Super. Pisa, Cl. Sci., IV Ser. 15, 115–147 (1988).

[Fe] Federer, H.: Geometric measure theory. Springer, 1969.

[FPU] Fermi, E., Pasta, J.R. Ulam, S.M.: Studies of nonlinear problems. In 'Collected works of E. Fermi', vol. 2 Chicago: Univ. Chicago Press, 1965.

[FSW] Fröhlich, J., Spencer, T., Wayne, C.E.: Localization in disordered nonlinear dynamical systems. J. Stat. Phys. 42, 247–274 (1986).

[Gr] Graff, S.M.: On the continuation of hyperbolic invariant tori for Hamiltonian systems. J. Differ. Equations 15, 1–69 (1974).

[GW] Guillemin, V., Weinstein, A.: Eigenvalues associated with a closed geodesic. Bull. Amer. Math. Soc. 82, 92–94 (1976).

[Her] Herman, M.: Inégalités a priori pour des tores Lagrangiens invariants par des difféomorphismes symplectiques. Publ. Math. IHES 70 (1989).

[HLP] Hardy, G.H., Littlewood, J.E., Polya, G.: Inequalities. Cambridge: Cambridge Univ. Press, 1967.

[HR] Helffer, B., Robert, D.: Asymptotique des niveaux d'energie pour des hamiltoniens à un degree de liberté. Duke Math. J., 49, 853–868 (1982).

[HS] Halmos, P., Sunder, V.: Bounded integral operators on L^2-spaces. Springer, 1978.

[K1] Kuksin, S.B.: Perturbation of quasiperiodic solutions of infinite-dimensional Hamiltonian systems. Izv. Akad. Nauk SSSR Ser. Mat. 52, 41–63 (1988); Math. USSR Izvestiya 32:1, 39–62 (1989).

[K2] Kuksin, S.B.: Hamiltonian perturbations of infinite-dimensional linear systems with an imaginary spectrum. Funts. Anal. Prilozh. 21:3, 22–37 (1987); Funct. Anal. Appl. 21, 192–205 (1987).

[K3] Kuksin, S.B.: Reducible variational equations and the perturbation of invariant tori of Hamiltonian systems. Matem. Zametki 45:5, 38–49 (1989); English transl. in Mathematical Notes 45:5, 373–381 (1989).

[K4] Kuksin, S.B.: An averaging theorem for distributed conservative systems and its applications to the von Karman equations. Prikl. Matem. Mekhan. 53:2, 196–205 (1988); P.M.M. USSR 53:2, 150–157 (1989).

[K5] Kuksin, S.B.: Perturbation theory for quasiperiodic solutions of infinite-dimensional Hamiltonian systems, and its applications to the Korteweg–de Vries equation. Matem. Sbornik 136 (178):3 (1988); Math. USSR Sbornik 64, 397–413 (1989).

[K6] Kuksin, S.B.: Quasiperiodic solutions of nearly integrable infinite-dimensional Hamiltonian systems. Preprint MPI/91-62 of Max–Planck–Institut für Mathematik, Bonn (1991).

[K7] Kuksin, S.B.: KAM-theory for partial differential equations. Preprint of ETH Zürich (1992). To appear in the Proceeding of The First European Congress of Mathematics (Paris, 1992).

[K8] Kuksin, S.B.: Perturbation theory for families of quasiperiodic solutions of infinite-dimensional Hamiltonian system and its applications. Dr.Sc.Thesis, LOMI, Leningrad, 1991 (Russian)

[K9] Kuksin, S.B.: On the inclusion of an analytic symplectomorphism close to an integrable one into a Hamiltonian flow. Preprint, ETH Zürich (1991) and Russian J. of Math. Physics 1:2 (1992).

[Ka] Kato, T.: Quasi-linear equations of evolutions, with applications to partial differential equations. Lecture Notes Math. 448, 25–70 (1975).

[Kap] Kappeler, T.: Fibration of the phase space for the Korteweg–de Vries equation. Ann. Inst. Fourier 41, 539–575 (1991).

[Kol] Kolmogorov, A.N.: On the conservation of conditionally periodic motions for a small change in Hamilton's function (in Russ.). Dokl. Acad. Nauk SSSR 98, 525–530 (1954); English transl. in Lecture Notes in Physics 93, 51–56 (1979).

[KP] Kuksin, S.B., Pöschel, J.: On the inclusion of analytic symplectic maps in analytic Hamiltonian flows and its applications. Preprint, ETH Zürich (1992). To appear in the Proceeding of the Conference on Dynamic Systems (St. Petersburg, October–November 1991), Basel: Birkhäuser, 1993.

[Kri] Krichever, I.M.: Perturbation theory in periodic problems for two-dimensional integrable systems, Sov. Sci. Rev. C. Math. Phys. 9, 1–101 (1991).

[Lax] Lax, P.D.: Development of singularities of solutions of nonlinear hyperbolic partial differential equations. J. Math. Phys. 5, 611–613 (1964).

[Laz] Lazutkin, V.F.: KAM-theory and semiclassical approximation to eigenfunctions. To appear in "Springer".

[Lio] Lions, J.-L.: Quelques methodes de resolution des problemes aux limites nonlineaires. Paris: Dunod, 1969.

[LL] Lichtenberg, A.J., Lieberman, M.A.: Regular and stochastic motions. Springer, 1983.

[LM] Lions, J.-L., Magenes, E.: Non-homogeneous boundary value problems and applications. Springer, 1972.

[Lo] Lochak, P.: Canonical perturbation theory via simultaneous approximation. Uspekhi Mat. Nauk 47:6 (1992); English transl. in Preprint of CMA ENS, Paris (1991), to appear in Russ. Math. Surv.

[Ma] Marchenko, V.A.: Sturm–Liouville operators and applications. Kiev: Naukova Dumka, 1977; English transl. Basel: Birkhäuser, 1986.

[McT] McKean, H.P., Trubowitz, E.: Hill's operator and hyperelliptic function theory in the presence of infinitely many branch points, Comm. Pure Appl. Math. 29, 143–226 (1976).

[Me1] Melnikov, V.K.: On some cases of conservation of conditionally periodic motions under a small change of the Hamiltonian function. Dokl. Akad. Nauk SSSR 165:6, 1245–1248 (1965); Sov. Math. Dokl. 6, 1592–1596 (1965).

[Me2] Melnikov, V.K.: A family of conditionally periodic solutions of a Hamiltonian system. Dokl. Akad. Nauk SSSR 181:3, 546–549 (1968); Sov. Math. Dokl. 9, 882–886 (1968).

[Mo] Moser, J.: Stable and random motions in dynamical systems. Princeton: Princeton Univ. Press, 1973.

[Mo1] Moser, J.: Convergent series expansions for quasi-periodic motions. Math. Ann. 169, 136–176 (1967).

[N] Nekhoroshev, N.N.: An exponential estimate of the time of stability of nearly-integrable Hamiltonian systems. Uspekhi Mat. Nauk 32:1, 5–66 (1977); Russ. Math. Surveys 32, 1–65 (1977).

[Nik] Nikolenko, N.N.: The method of Poincaré normal forms in problems of integrability of equations of evolution type. Uspekhi Mat. Nauk 41:5, 109–152 (1986); Russ. Math. Surveys 41:5, 63–114 (1986).

[NMPZ] Novikov, S.P., Manakov, S.V., Pitaevskij, L.P., Zakharov, V.E.: Theory of solitons. Nauka, Moscow, 1980; English transl. Plenum Press, New York, 1984.

[P1] Pöschel, J.: On elliptic lower dimensional tori in Hamiltonian systems. Math. Z. 202, 559–608 (1989).

[P2] Pöschel, J.: Small divisors with spatial structure in infinite dimensional Hamiltonian systems. Comm. Math. Phys. 127, 351–393 (1990).

[P3] Pöschel, J.: Integrability of Hamiltonian systems on Cantor sets. Comm. Pure Appl. Math. 35, 653–695 (1982).

[P4] Pöschel, J.: On Nekhoroshev estimate for quasi-convex hamiltonian systems. Math. Z. (to appear).

[PT] Pöschel, J., Trubowitz, E.: Inverse spectral theory. Boston: Academic Press, 1987.

[RS] Reed, M., Simon, B.: Methods of modern mathematical physics, Vol. 1–4. Academic Press, 1980.

[Ru] Rüssmann, H.: Über invariante Kurven differenzierbarer Abbildungen eines Kreisringes. Nachr. Akad. Wiss., Göttingen II, Math. Phys. Kl., 67–105 (1970).

[Sev] Sevryuk, M.B.: Invariant m-dimensional tori of reversible systems with a phase space of dimension greater than $2m$. J. Sov. Math. 51, 2374–2386 (1990).

[SG] Shilov, G.E., Gurevich, B.L.: Integral, measure and derivative: a unified approach. Dover Publications, Inc., 1977.

[SM] Siegel, C.L., Moser, J.: Lectures on celestial mechanics. Springer, 1971.

[SZ] Salamon, D., Zehnder, E.: KAM theory in configuration space. Comment. Math. Helv. 64, 84–132 (1989).

[U] Ulam, S.: Problems in modern mathematics. John Wiley and Sons, Inc., 1964.

[VB] Vittot, M., Bellissard, J.: Invariant tori for an infinite lattice of coupled classical rotators. Preprint CPT-Marseille, 1985.

[W1] Wayne, C.E.: Periodic and quasi-periodic solutions of nonlinear wave equations via KAM theory. Commun. Math. Phys. 127, 479–528 (1990).

[W2] Wayne, C.E.: The KAM theory of systems with short range interactions, 1 and 2. Comm. Math. Phys. 96, 311–329 (1984); 331–344 (1984).

[W3] Wayne, C.E.: Bounds on the trajectories of a system of weakly coupled rotators. Comm. Math. Phys. 104, 21–36 (1986).

[War] Ware, B.: Infinite-dimensional versions of two theorems of Carl Ziegel. Bull. Amer. Math. Soc. 82, 613–615 (1976).

[Z1] Zehnder, E.: Generalized implicit function theorem with applications to some small divisor problems, I and II. Commun. Pure Appl. Math. 28, 91–140 (1975); 29, 49–111 (1976).

[Z2] Zehnder, E.: Siegel's linearization theorem in infinite dimensions. Manus. Math. 23, 363–371 (1978).

[ZIS] Zakharov, V.E., Ivanov, M.F., Shur, L.N.: On the abnormally slow stochastisation in some two-dimensional field theory models. Pis'ma. Zh. Eksp. Teor. Fiz. 30:1, 39–44 (1979); JETP Letters 30:1 (1970).

INDEX

Printing: Weihert-Druck GmbH, Darmstadt
Binding: Buchbinderei Schäffer, Grünstadt